Do-It-Yourself
Insulation &
Weatherstripping

By the Editors of Sunset Books and Sunset Magazine

Lane Publishing Co. • Menlo Park, California

Because saving energy is of national importance, federal and state governments are pitching in to make energy-oriented home improvements easier for homeowners. To find out about federal support, call the nearest office of the Energy Research and Development Administration. And, many states have incentive measures worth exploring. For information about state support, try your state's public utilities commission (some have an energy conservation unit). Also check with your utility company. Some utilities offer services ranging from free information and professional advice to energy-improvement loans.

Supervising Editor: Donald W. Vandervort

Research and Text: Jim Barrett, Mike Scofield, William M. Heineken, Tony Di Angelo

Design and Illustrations: Joe Seney

Cover: Photographed by Norman A. Plate

Editor, Sunset Books: David E. Clark

Fifth Printing December 1980

Contents

Keeping your energy costs in line

As our energy resources dwindle, costs are rising. Gas and electricity are becoming too scarce, too precious, too expensive to waste. So how do we respond to the challenge? We make the most of what we have—we conserve. By insulating and weatherstripping our homes, we keep them comfortably heated or cooled with less fuel, reducing our heating bills and saving valuable resources. In short, we get more for less.

Insulation and weatherstripping help keep unwanted weather outside. A properly insulated and weatherstripped house is comfortable and energy-efficient the year around. In the winter, the heat remains inside and the cold stays out. And in the hot summer, cool air is captured inside where the outside heat cannot enter.

In addition to saving energy resources and money, insulation and weatherstripping can reclaim areas—even entire rooms—that would otherwise be practically useless because of heat gain or loss. In winter, they can eliminate chilling drafts that blow through the basement, sweep across floors, or engulf window and door areas. Insulation and weatherstripping make your house more usable and comfortable by helping to regulate the climate inside.

Where your house leaks heat

When your heating system is in operation, heat flows through the house. Eventually it dissipates and is exchanged for cool air. The speed of this process depends strongly on how complete the barriers are between the warm air inside your house and the cold air outside. In summer, the hot air is outside and the cool air inside, but the principle is the same.

Unless your home was built within the past few years, it probably isn't properly insulated or weatherstripped. The lack of adequate insulation and weatherstripping accounts for excessive heat losses or gains.

Drawing **5-A** shows where most of the heat is lost from a typical two-story house during the winter. In all, about 60 percent escapes through the roof, walls, and floor; this is where insulation can cut heat loss by from one-third to one-half. Approximately one-quarter of the heat loss is through windows and doors, depending upon their size and number. Weatherstripping and storm doors and windows can cut this figure by 50 percent. Another 16 percent of the heat escapes by infiltration through vents, chimneys, and similar openings. Some of this loss is inevitable and in fact necessary. But some can be eliminated if you understand the facts about proper ventilation.

The ratio of heat loss through roof, walls, and floors changes drastically in a one-story house, where less of the house is walls and more is roof and floor. In a typical one-story house, about 28 percent of the heat escapes through the roof (see drawing **5-A**). On the average, a one-story house requires about 15 percent more energy to heat than a two-story house of the same size.

This book discusses all these areas of heat loss (or gain). The section on insulation begins on page 8. The weatherstripping section starts on page 50. You'll find information on storm doors and windows on page 76. A special feature on ventilation is on page 22. In addition, you'll find helpful household tips for saving energy on page 49.

5-A

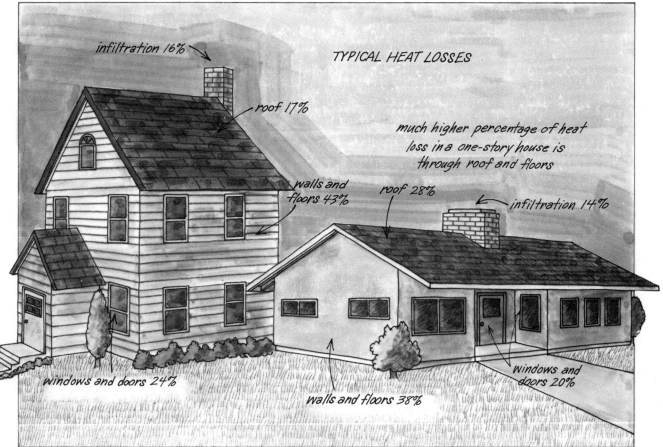

TYPICAL HEAT LOSSES

infiltration 16%

roof 17%

much higher percentage of heat loss in a one-story house is through roof and floors

walls and floors 43%

roof 28%

infiltration 14%

windows and doors 24%

windows and doors 20%

walls and floors 38%

Based on uninsulated houses of similar construction with equal floor, window, and door area.

How heat moves

Because insulation and weatherstripping are designed to control the movement of heat, let's take a closer look at how heat travels.

Heat always moves from warmer to colder areas—it seeks a balance. If the interior of your house is colder than the outside air, the house draws heat in from outdoors. Under the opposite conditions, the house gives off heat. The greater the temperature difference, the faster the heat flows to the colder area.

Heat moves from one place to another in three ways: conduction, radiation, and convection. It travels through house walls by all three methods.

Conduction. Heat travels through a solid object, or from one solid object to an adjoining one, by creeping through the material from one molecule to another (see drawing **6-A**). Heat penetrates

6-B

RADIATION OF HEAT

is by radiation that the sun's heat warms walls and roofs even on a cool day. See drawing **6-B**.

Convection. Through large air spaces, molecules of air can transfer heat from warm surfaces to cold ones. Heated air rises (drawing **6-C**). In an enclosed space, heat forms convection currents: air warms and rises, then (as it gives off heat to

6-A

HEAT'S MOVEMENT BY CONDUCTION

building materials in this way. In general, the denser a material is, the more quickly heat moves through it. (Some metals are exceptions to this rule, however.)

Radiation. Wave motion carries heat in much the same way that it transmits light. By radiation, heat can jump directly from warmer objects to colder ones without warming the air between them. At the surface of the colder object, part of the energy is absorbed; some may be reflected. It

6-C

HEAT'S MOVEMENT BY CONVECTION

the surrounding surfaces) it cools and sinks again. When this occurs in spaces between structural framing members in a ceiling or wall, a substantial amount of heat may be transferred and lost. But in small spaces or cavities, where convection currents can't develop, air is a good insulator.

Movement of heat through a wall. Heat travels through a typical wall or roof by all three of these methods (see drawing **7-A**). The warm air in a room moves to interior wall surfaces by convection and radiation. Then it travels through the interior wall surface by conduction. It gets from the inner wall surface to the exterior wall surface by all three means—by convection currents through the space between framing members, by radiation from one surface to another, and by conduction through the framing materials. It leaves the exterior wall surface by convection and radiation.

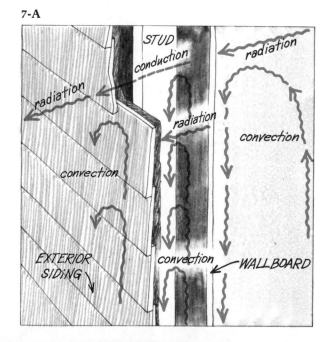

7-A

How our bodies stay comfortable

Our bodies act like miniature furnaces, continuously throwing off surplus heat. The amount of heat we produce varies with our age and activities. A young person throws off more body heat than an elderly one, and anyone gives off more heat when walking than when sitting, more when running than when walking.

Heat always moves from a warm surface to a colder one. Whether we feel warm or cold depends upon whether our surroundings are warmer or colder than we are.

Actually, our bodies constantly produce more heat than they need. For comfort, they must be able to transfer this heat to surrounding objects, or to the air, at a balanced rate. If we don't lose heat fast enough, we feel hot. If we lose it too fast, we feel cold. In between, we're comfortable.

If we are to be comfortable, our skin temperature must remain relatively constant whether we are active or resting, whether it is winter or summer. Our clothes are designed to maintain the first phase of balance—our houses should complete the process.

In a house whose exterior surfaces are insulated and weatherstripped, we lose less body heat to those surfaces because they are warmer. Because we lose less body heat, we feel more comfortable and cozy in a house that is cooler. A house that is properly "buttoned up" is comfortable at 65° to 68°. A poorly weatherstripped, inadequately insulated house requires a temperature of about 3° more (see drawing **7-B**).

7-B

Insulation

Keeping heat inside your house during winter and outside it in summer is what insulation is all about.

Suppose your furnace is in good repair and your doors and windows are weatherstripped and tight. If on a chilly day you can still feel a miniature Niagara of cold air falling down the walls and cascading across the floor, your house may be poorly insulated—or not insulated at all.

How much energy will you save by installing insulation from scratch or beefing up your existing insulation? That depends on many variables: the amount of insulation your house already has, your ceiling height, the number and size of windows, the presence or absence of storm doors and windows, the efficiency of your heating system, orientation of the house to the sun, and, of course, the weather conditions where you live. The exact amount of money you'll save also depends on the cost of fuel in your area.

You can save more than one-third of your house's total heat loss by insulating walls and roof. In most cases, fuel savings pay back the cost of insulating in less than 5 years. Because insulation is a permanent improvement, your energy savings will continue through the years you own the house. And when you sell, the insulation will increase the market value of your house.

Where insulation belongs

Insulation belongs inside any barrier between a heated space and an unheated one. Applied to the structure of a house, this means that insulation should be in all exterior walls, in attics, under floors exposed to the outside, and on heated basement walls. Drawing **10-A** illustrates this.

Spots in which insulation is often overlooked include the wall between living space and an unheated utility room, storage room, or garage; dormer walls and ceilings; exterior walls between levels in a split-level home; knee walls next to heated attic rooms; overhead collar beams in a heated attic; and floors over vented crawl spaces, over unheated spaces, and cantilevered out over exterior walls.

Insulation should form a complete envelope around all living areas of your house, leaving no openings except doors, windows, and necessary venting. For further conservation, you should insulate around heating ducts and pipes (see page 48 for information on insulating ducts and pipes).

A priority list

Here are some factors to consider when deciding which areas to insulate first.

Attic. Particularly in a one-story house, most heat leaves through the attic. By installing sufficient attic insulation where there is none, you can cut up to 30 percent of your fuel bill. If your attic is unfinished, insulation is relatively easy to install; put this at the top of your priority list.

If you have a finished attic, insulate as much of it as you can (see page 34). Flat roofs and mansard roofs present special access problems; talk to a contractor if you're considering insulation for one of these.

Walls. Exterior walls with no wall surface on one side are easy to insulate, and this should be a priority task. But most walls are covered on both sides, and for those you must either remove the wall covering on one side, then install rigid-board insulation over it and recover, or hire a professional to blow in or foam in insulation through scores of holes drilled for access. Removing the wall coverings would be absurd unless you plan to remodel the house.

As for hiring a contractor to insulate finished walls, you'll have to weigh the savings against the cost. Tests done by the Energy Research and

Development Agency show that blowing in insulation doesn't do a complete job, and foaming in insulation is often too expensive to pay for itself in a reasonable period. But get a few estimates to find out the exact cost (see page 19 for information on working with contractors). By insulating walls properly, you can save 16 to 20 percent of your heating costs.

Basement or crawl space. If reasonably accessible, this is a good area to insulate. Doing so can save 5 to 15 percent of your heating costs. It is generally easy to insulate under floors that have an unfinished basement or crawl space below.

10-A

ceiling joists

short exterior walls

wall to unheated garage

If your basement is heated and its walls are concrete, insulating involves some carpentry work. To install batts or blankets, you must build out each concrete wall with wooden framing members to provide room for them. Rigid-board insulation is attached to wood nailing strips bolted to the wall. After installing insulation, you cover it with gypsum wallboard or another approved wall covering.

In deciding whether or not to insulate a concrete basement wall, you must again weigh your projected savings against the cost and work of doing the job. If you plan to finish the basement anyway, insulating would be a wise investment.

Pipes, ducts, and water heater. Insulating these, as discussed on page 48, is an easy, low-cost job that returns sizable energy savings.

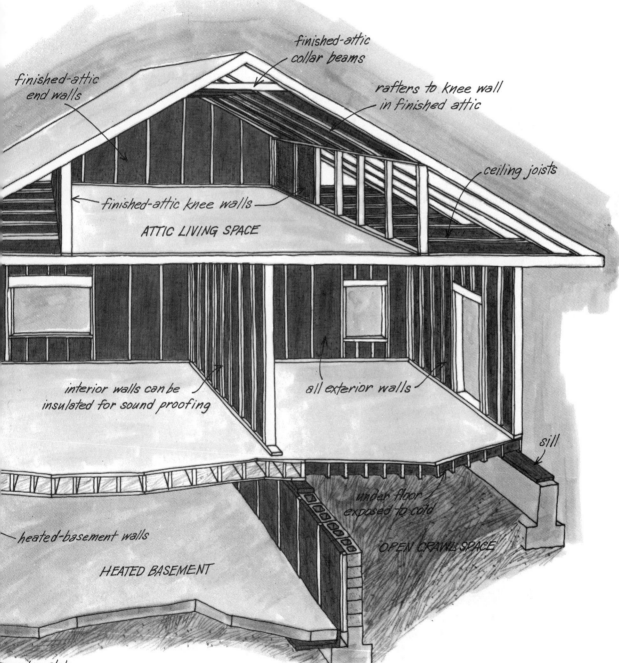

finished-attic
collar beams

finished-attic
end walls

rafters to knee wall
in finished attic

ceiling joists

finished-attic knee walls

ATTIC LIVING SPACE

interior walls can be
insulated for sound proofing

all exterior walls

sill

heated-basement walls

under floor
exposed to cold

OPEN CRAWL SPACE

HEATED BASEMENT

under slab

How much is enough?

The right amount of insulation to install depends upon two main factors: 1) the amount of insulation already in your house, and 2) the R-values (insulative values) you want the walls, ceiling, and floor to have. This section will tell you how to check for existing insulation, explain what R-values are, discuss typical amounts of insulation, and show you how to figure the right quantities to buy.

Checking for existing insulation

If your house has an unfinished attic, look for insulation between the joists that frame the attic floor. Put on some old clothes and set up a stepladder under the attic access hole (it's probably in a closet or hallway). Taking a portable light and measuring tape, climb up inside. Watch out for nails protruding through the roof sheathing overhead. **Do not step on the ceiling board between the joists; it won't support you.** Standing or kneeling with one foot or knee on each of two joists, or on boards used as temporary flooring, survey the situation.

Measure the length and width of the attic so you can calculate square footage later. Also, for reference when buying insulation, measure the distance between typical joists.

If you find any existing insulation between the joists, measure its thickness. Be sure it's dry, not soggy. Also check to see if there is a vapor barrier of foil, polyethylene, or building felt underneath the insulation. If there is one, try to find out if it is in good shape (not torn).

If your attic isn't insulated, chances are good that your walls and floors aren't insulated either.

Looking for existing insulation inside an exterior wall that is covered on both sides involves more detective work. After turning off the power, remove a switch cover, electrical outlet cover, or light fixture (see drawing **12-A**). Look-

12-A

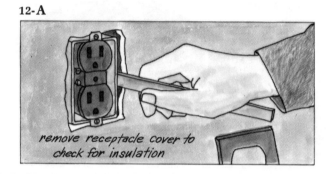

remove receptacle cover to check for insulation

ing inside the wall through the resulting opening, check for insulation between walls studs. You may need a flashlight for this. If possible,

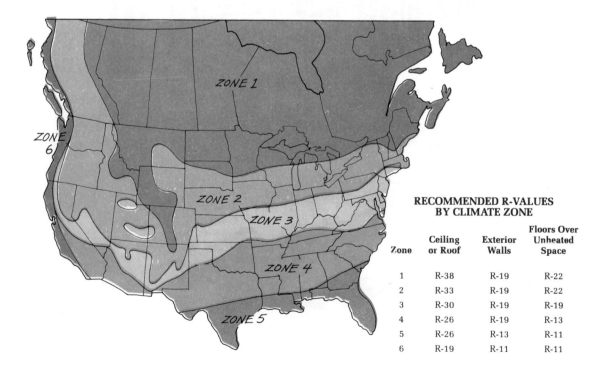

RECOMMENDED R-VALUES BY CLIMATE ZONE

Zone	Ceiling or Roof	Exterior Walls	Floors Over Unheated Space
1	R-38	R-19	R-22
2	R-33	R-19	R-22
3	R-30	R-19	R-19
4	R-26	R-19	R-13
5	R-26	R-13	R-11
6	R-19	R-11	R-11

measure the thickness of any insulation you see. Take care how you measure—it may be bunched up. Replace the fixture or cover, then turn the power back on.

If this kind of inspection leaves you unsure, you can place a thermometer on the interior side of an exterior wall. With the room at 68° and the outside temperature about 50°, a well-insulated wall should register about 65°. If it's much lower than 65°, the wall is probably not insulated. Of course, this method is not as reliable as actually looking for insulation.

Under a floor, you can generally see insulation simply by going down to the basement or into a crawl space under the house.

After you've checked all barriers that face the outside or unheated spaces and noted the existence of any insulation, convert the thickness of existing insulation into R-values by using the chart below.

THICKNESSES REQUIRED FOR SPECIFIC R-VALUES

	Blankets, Batts		Loose Fill (Blown) Minimum		
R-Value	Fiberglass	Rock wool	Fiberglass	Rock wool	Cellulose
R-11	3½"	3"	5"	3¾"	3"
R-19	6"	5¼"	8¾"	6½"	5"
R-22	6½"	6"	10"	7½"	6"
R-30	9½"*	9"*	13¾"	10¼"	8"
R-38	12"*	10½"*	17½"	13"	10½"

*Requires two blankets or batts

What is an R-value?

Insulation amounts were once measured by thickness, in inches. But because different types of insulation require varying thicknesses to produce the same resistance to heat flow, inch ratings were sometimes misleading.

Insulation is now rated by its resistance to heat flow. This rating is called an "R-value." Think of R as abbreviating *resistance*. The higher the R-value a material has, the better its resistance to heat flow.

Even though one roll of insulation might be thicker than another, if they're both marked with the same R-value, they'll be equally effective as insulation.

But keep one thing in mind: the R-value shown on an insulation package rates the material's performance when properly installed. If insulation is packed too tightly, installed incompletely, or allowed to get wet, its R-value will be reduced. Another tip: be suspicious of any insulation package that isn't marked plainly with an R-value.

What are average amounts?

As an average, the National Bureau of Standards recommends R-11 insulation in exterior walls and floors and R-19 in the ceiling or roof. For walls between heated and unheated rooms—in a basement, for example—the recommendation is for R-11. The insulation industry claims that a house meeting these standards needs less than 50 percent of the energy required to heat an uninsulated house.

But average may not be right for your house. Higher R-values may be necessary if the climate you live in is colder—or warmer—than average. Also keep in mind that energy costs are going up; it may pay in the long run to install insulation now with higher R-values than average.

Insulation standards vary throughout the country. Of course, many existing houses have no insulation at all. In most states, only new homes are presently required to meet certain insulation standards.

What is your climate?

To determine the proper insulation values for your home, consider the map on page 12. It divides the United States and part of Canada into six climate zones. (Alaska is in zone 1; Hawaii is in zone 6.) Find the zone in which you live. Then check the chart to get the proper R-values for your zone. Use these R-values when choosing your insulation.

How to figure your needs

Before you buy insulation, you'll need to figure out how much to get. All you must do for this is take a few measurements and do some basic multiplication. Here's how to figure the necessary amount for each area:

1) Find the square footage of the area to be insulated by multiplying the length times the width (in feet).

2) Next, compute the total square footage of doors, windows, and any other areas to be excluded. Subtract this figure from the one found in step 1.

3) If studs or joists are spaced on 16-inch centers, multiply the total square footage by 0.9 to allow for the area taken up by studs or joists. If studs or joists are on 24-inch centers, multiply by 0.94.

4) The resulting figure is the total square footage of insulation you need. Each bag or bundle should be labeled with its square-foot coverage.

Types of insulation materials

There is no mystery about insulation. Each of the materials that make up your house has some insulative value. But their effectiveness varies greatly. A 1-inch blanket of insulation has the same insulative value as 3½ inches of pine planking, 22 inches of common brick, 40 inches of concrete, or 54 inches of stone. There is actually a measurement for the heat conductivity of combined building materials—it's called a "U-value"—but the calculations required are very complex. It's better to stick to R-values.

As mentioned earlier, air is a good insulator when confined to a small space or to cells inside a porous material. Of course, dense materials such as stone have few, if any, air spaces. On the other hand, fibrous or porous materials contain plenty of air. These materials are the most common types of insulation. They break up the air space between exterior and interior house walls, floors, and ceilings.

Where heat gain or loss through radiation is significant, foil is used to reflect the heat back toward its source. Most commonly, it is provided as a facing on blanket-type fibrous insulation (it serves double duty as a vapor barrier; see page 24). In some cases, foil is a backing on gypsum wallboard. And in some areas, foil alone is used as insulation (see "Reflective foil," page 18).

Insulation is manufactured from several materials, all of them highly porous. They include fibers or granules of glass, mineral, and organic materials, as well as some types of plastics. A good insulation product is resistant to vermin and moisture and to any change that would hinder its effectiveness. And it is fireproof when properly applied.

The chart on the next page compares insulation materials. These R-values should not be used for estimating thickness. The actual values for a given insulation change with temperature and, in the case of some materials, depend on the manufacturing process used.

Of course, insulative value isn't the only factor to consider when choosing a material. Cost, method of installation, and general traits all count—and these factors vary greatly. Here is a brief description of each type.

Vermiculite and perlite. These are not as commonly available as other types of insulation. Vermiculite is made from expanded mica. It is packaged in loose-fill granules (the smallest available) that are hand poured into attics and hard-to-reach spots. Though it is relatively easy to install, vermiculite tends to absorb moisture, becoming mushy and decreasing its already-low

insulative value. Perlite, a type of volcanic rock, is comparable to vermiculite but has a slightly higher R-value.

Mineral wool. Fiberglass and rock wool, the two products grouped under this heading, make up 90 percent of homeowner-installed insulation. Fiberglass is made from glass fiber. Rock wool, spun from molten slag rock, has a slightly higher R-value than fiberglass.

Mineral wool is available in several forms: packed into flexible blankets and batts, shredded for hand pouring or machine blowing, and compressed into rigid boards.

Though both fiberglass and rock wool are irritating to skin, rock wool is the less bothersome of the two.

Polystyrene. A type of plastic, polystyrene is a rigid-board material. Though it dents easily and is highly combustible, its weather and moisture resistance make it excellent for below-grade or exterior-wall applications. Used indoors, it is a fire hazard unless covered by ½-inch gypsum wallboard. In some cases, polystyrene should be installed by a qualified contractor.

Cellulose. Made from recycled paper products, cellulose is a shredded insulation that is normally machine blown into unfinished attics by a contractor or by a homeowner using a dealer-loaned machine.

Because of its fine consistency, cellulose can also be blown into walls that are covered on both sides—but this should be done only by a qualified insulation contractor. Even when a contractor performs the work, the coverage may be somewhat incomplete.

If you can't get a machine for blowing cellulose into an attic, you can spread it by hand. Though it is not irritating to skin, cellulose is quite dusty. It is flammable unless treated, so be sure the bags you buy are clearly marked as meeting federal specifications. Also make sure the material has been treated for rodent resistance.

Because cellulose fibers have natural air cells inside them, cellulose insulation has relatively high R-values.

Urea-formaldehyde foam. Though quite expensive, urea formaldehyde is excellent for insulating walls finished on both sides. It fills wall cavities totally and, once inside, creates its own vapor barrier (see page 24). Because the equipment and procedures are specialized, a contractor must install it. It has excellent resistance to fire and has high R-values. On the negative side, it can cause a slight odor that lingers for a week or two, and the R-values may tend to decrease slightly with time. For more about this material, see the section on insulating walls, page 39.

Urethane. Though urethane has the highest R-values of all insulations, it gives off poisonous gases when ignited. It comes in rigid-board panels or as a foam like urea-formaldehyde. If foamed in place by a contractor, it completely fills the spaces where it is installed and it creates its own vapor barrier. But because of the toxic gases it may give off, urethane is not recommended for use in walls.

In some localities, urethane foam is applied to roof surfaces by contractors. Like rigid boards, this relatively new insulating method can be an excellent way to insulate houses with no attics.

COMPARISON OF INSULATION MATERIALS

TYPE OF INSULATION	FORM	APPROX. R-VALUE PER INCH OF THICKNESS	RELATIVE COST (1=least; 5=most)	WHERE USED
Vermiculite	Loose fill (poured)	2.1	4	Poured into small spaces. Sometimes poured between ceiling joists.
Perlite	Loose fill (poured)	2.7	4	Poured into small spaces. Especially useful for hollows in concrete blocks.
Fiberglass	Blankets and batts	3.1	1	Fitted and secured in open framing in walls, ceilings, and floors.
	Loose fill	2.2 (blown)	1	Poured or blown into spaces in attic floor framing. Blowing wool is contractor installed.
	Boards	4.5	5	Attached to wall or ceiling surfaces.
Rock wool	Batts	3.7	1	Fitted and secured in open framing in walls, ceilings, and floors.
	Loose fill	2.9 (blown)	1	Poured or blown into spaces in attic floor framing. Blowing wool is contractor installed.
Polystyrene	Boards	3.5–5.4	5	Attached to exterior surfaces. Usually contractor applied.
Cellulose	Loose fill	3.6 (blown)	2	Blown into attics and some walls.
Urea-formaldehyde	Foam	4.2	5	Contractor installed in enclosed framing cavities.
Urethane	Foam Boards	4.5	5	Contractor installed in some framing cavities, on top of roofs, and on other exterior surfaces.

The highest R-values are obtained with the first 3″ of thickness. Value per inch decreases on a sliding scale as thickness increases.

How insulation is packaged

Homeowner-installed insulations are manufactured and packaged according to the way they're meant to be installed. When selecting a type of insulation, you'll need to distinguish among blankets, batts, loose fill, rigid boards, and foil. And within these categories you'll have to make further choices. Here is a brief rundown to acquaint you with the most common forms of homeowner-installed insulations and their uses.

Blankets. Mineral-wool insulation is available in large rolls properly called "blankets." Blankets are used for insulating unfinished attic floors or rafters, unfinished walls, and the undersides of floors.

Blankets are preferred over batts (see below) for insulating long runs between joists or rafters, or for walls that are not of standard 8-foot height.

The most common thicknesses of blanket insulation are 3½ inches (R-11) or 6 inches (R-19). Length of the 3½-inch blanket is usually 56 feet; a common 6-inch blanket length is 32 feet.

The width of blankets depends upon the way they are to be installed. Some are made to be wedged between wall studs or ceiling joists without fasteners; these come in 15¼-inch and 23¼-inch widths. Other blankets are faced on one side with foil or kraft paper that is to be stapled to sides of studs and joists. Foil and kraft-faced blankets are available in 15-inch and 23-inch widths. All are sized to fit standard wall-stud and joist spacings.

Foil or kraft paper facings applied to blankets serve as vapor barriers (see page 24). Foil can also reflect a small amount of radiant heat back into the room.

Unless you install a separate vapor barrier, unfaced blankets are best used only where a vapor barrier already exists, where no vapor barrier is necessary, or where you plan to add an overall vapor barrier of 2-mil polyethylene sheeting or foil-backed gypsum wallboard.

Choosing between kraft and foil for a vapor barrier depends on the job. Foil makes a slightly tighter vapor barrier (0.5 perm or less compared

BLANKET

BATTS

Fiberglass (foil faced)

Fiberglass
(foil faced)

Rock Wool
(kraft-paper faced)

to 1 perm or less for kraft paper). Kraft paper doesn't tear as easily as foil but is easier to cut. And foil stays stapled in place better.

Blankets 1 inch thick are available for sealing the sill of a wall where it meets the foundation; this job is done during construction. To insulate heating ducts, you can also get 1-inch or 2-inch-thick blankets that are about a foot wide.

Batts. Mineral-wool insulation also comes in batts. These are simply blankets that have been precut to lengths of 4 or 8 feet. Batts are more convenient than blankets for insulating rooms with standard 8-foot ceilings, for insulating framing runs that are in increments of 4 feet, and for use where elbow room is limited.

Batts come in the same thicknesses and widths as blankets. Rock wool batts are slightly thinner than fiberglass batts or blankets with the same R-values.

You can buy batts in several forms: unfaced, faced on a single side with a kraft paper or foil vapor barrier, or faced both sides with foil or kraft paper. The double-faced type has a vapor barrier on one side and perforated "breather paper" on the other; it's popular for insulating under floors, where fibers would otherwise tend

to drift down into the installer's face. The breather paper should be stripped off after installation and not left exposed, because it is flammable.

Loose fill. Though loose-fill insulation is best used in unfinished attics, some types are blown into walls that are covered on both sides. In an attic, loose fill can be built up to any thickness to achieve practically any desired R-value.

The main problem with loose fill is that the insulation provides no vapor barrier. If one is needed, as it often is, you must install or create a separate vapor barrier (see page 24).

Loose-fill insulation is of two types: granules and fibers.

Granular insulations are poured in place and spread by hand. They fill tight spots more completely than fibrous insulations. For more about them, see "Vermiculite and perlite," page 14.

Fibrous insulations are more commonly used. They are either poured or blown into place. Two kinds, loose-fill fiberglass and rock wool, can be poured by hand, but are usually machine blown into place by a contractor. They are best used in attics; when installed in walls, they tend to hang up on nails and other obstructions, and their

LOOSE FILL

Pouring Wool

Blowing Wool

Vermiculite

Cellulose

effectiveness is reduced by this incomplete coverage.

Bags of fiberglass and rock wool are labeled either "pouring wool" or "blowing wool." If you're going to hand pour the material, be sure to get pouring wool.

Cellulose is sometimes hand poured, but machine installation is much better and easier. This is the only type of insulation that a homeowner can blow in; most dealers of cellulose loan a pneumatic machine rent-free for this job. Blowing in cellulose is one of the easiest ways to insulate an unfinished attic.

Some contractors blow cellulose into walls. It loads faster into the blowing machine than blowing wools and doesn't settle in walls as quickly. But like blowing wools, it doesn't completely fill the cavities.

Hand pouring loose fill is difficult. You must carry the bags of material up to the attic and, in hard-to-reach areas, do a lot of hunching over. To control the R-value, you must pull apart and fluff the material to approximately three times its packed volume. This fluffing can be done with a garden rake once the insulation is in place, but the most effective way to do it is by shaking the material in a closed box. No matter how you do it, the R-value is difficult to keep constant.

Rigid-board panels. Also referred to as insulative sheathing, rigid-board panels are fast becoming a popular do-it-yourself alternative to conventional blanket, batt, and loose-fill insulation. Though primarily used in new construction, rigid-board panels can also be installed in existing homes, either in lieu of or in addition to conventional insulation materials in almost any location.

In new construction, rigid boards are used as sheathing under exterior siding, as underlayment for roofs, and around foundations. Used in tandem with conventional insulation, rigid boards provide increased R-values necessary to meet the new requirements and standards of federal agencies, utility companies, and building codes. In existing houses, rigid boards are easily applied over exterior siding and interior walls. Also, this type of insulation is often the best answer for insulating open-beam ceilings and masonry basement walls.

Rigid-board panels are available in 2 and 4-foot widths in a variety of lengths; facings include building felt, kraft paper, and foil on one or both sides. Four basic types are available:

Asphalt-impregnated fiberboard, the oldest type, has the lowest R-value of all the rigid-board panels, making it less popular than the others.

Beadboard, made by fusing together polystyrene beads (similar to styrene packing materials), has an R-value higher than fiberboard, but lower than rigid boards (below). Like rigid foam boards, beadboard is easily dented and must be handled carefully.

Fiberglass sheathing is made by compressing fiberglass wool between a tough facing material, forming semi-rigid boards. Its R-value is equivlalent to beadboard. A vapor barrier is required where applicable.

Rigid-foam boards have the highest R-values of all rigid-board insulation. They are produced from sophisticated chemical formulations (based on polyurethane or polystyrene) and foamed into rigid sheets. Rigid-foam boards form their own vapor barrier. This material dents easily and must be handled carefully.

Because of their high flammability, most rigid-board panels must be covered by ½-inch gypsum board on interior walls and ceilings. Installation methods vary among the various types of rigid-board panels. For this reason, installation is not covered in this book; most manufacturers provide installation details with their products, though.

Reflective foil. Unlike the other insulations discussed here, foil does not use the principle of entrapped air to slow heat flow. It gets its insulative value by combining enclosed air spaces with reflective surfaces.

The most common type comes packaged in corrugations backed with kraft paper that open like an accordion prior to installation.

The insulative value of reflective air spaces varies widely, depending upon the direction of heat flow. Reflective insulations are most effective when heat flow is downward. For this reason, reflective insulations can work effectively when used in a floor but may not be satisfactory in walls or ceilings, where heat moves horizontally and upward.

Reflective foils are designed to control radiant heat movement only, so their R-values are low. And they must be installed so that they face an air space. If they are squeezed between two surfaces with no air space on either side, they have no insulative value. To be effective they must be installed carefully, without rips or tears.

Rigid-board panels applied as sheathing *insulate walls.*

How to work with a contractor

If, for one reason or another, you've decided to hire a contractor to do your insulating, you'll want to find a qualified one for the job. And you will probably want to be sure you're getting your money's worth.

This page will tell you how to find contractors, choose one, specify work in a contract, and make sure the job you receive is what you paid for.

Finding a contractor

The best way to find a contractor is by recommendation from someone whose judgment you trust: a neighbor, friend, or relative who has had similar work done.

Or you can ask for a recommendation from your utility company, a local office of the National Association of Home Builders, the Home Builders Association, or a government-funded, nonprofit home improvement program, if your area has one.

The Yellow Pages are a source too. Look under "Insulation Contractors—Cold and Heat." The size of the company you deal with is irrelevant if all you want is a good job at the right price.

Choosing a contractor

Try to gather the names of three or four different contractors—the more you have to choose from, the better your chances of getting the job done when you want it done. And if each knows that others are bidding on the job, your chances are better of getting a good price.

Once you get bids, check up on the qualifications of the contractor who is most likely to do the work. Tell him you want to get references from a few of his past customers, and ask him for their phone numbers or addresses. Then contact some of these people to see if they were satisfied with the work.

If you want to go one step further, check your local Better Business Bureau to see if any complaints have been lodged against the contractor or his business.

Read the sections in this book that discuss the job you have in mind to see what is involved. Try to get a clear idea of what the job entails. To get bids that you can compare, be sure to give each contractor the same list of specifications: location of the insulation, type of insulation, R-value desired, installation method, and extent of work (who will patch up, clean up, dispose of refuse, and so forth).

Don't select a contractor by price alone; try to get a feeling for the quality of work each will do.

The contract

Once you've chosen a contractor, have him put in a contract all the specifications of the bid, plus the cost, method of payment, and his warranty. Not only does a contract bind both parties, it defines responsibilities so there are no misunderstandings or hard feelings after the job is done. To make the contract valid, both of you must sign it and retain copies.

Limit your down payment to 25 percent. And find out if the contractor's insurance covers his workers for injuries and your house for damage. If it doesn't, better check your homeowner's policy.

Checking for quality

If the insulating will be done with loose fill, ask the contractor to show you one of the bags of material he plans to install. A chart on the label should give the square footage the bag will cover for the desired R-value. (See page 13 for information on figuring how much insulation is needed.) Use the figures on the label and the square footage to be insulated to compute the number of bags the contractor will need to do the job right.

You can either keep tabs as he installs the material or ask him to save the empty bags for your count. If the number of bags installed is short of your calculations by more than four or five, you should talk with the contractor about the discrepancy.

For foam installations, there is no easy way to keep tabs on the amount installed. Ask for written certification of the R-value achieved.

No matter what type of insulation is installed, get a certificate identifying the type of insulation, the manufacturer's name, and the R-value installed. Keep it on file for reference when you sell your house.

If batts or blankets are installed, be sure all areas are covered completely; if vapor barriers are part of the deal, check to see that they are intact.

If, after reading this, you are still concerned about quality, call your utility company. In some localities, they will send an inspector out to check the job.

Tools and supplies

Most insulation work requires relatively few tools and supplies. Insulation materials are designed to go in easily, without much fuss. Here are some suggestions for equipment to have on hand (see drawing **20-A**).

Protective gear. Because insulation materials are either dusty or irritating to skin, wear gloves when working with them. Cotton gloves give you protection yet don't cause your hands to perspire. And, for installing fiberglass or rock wool, wear loose-fitting clothes that won't work mineral fibers into your skin. A cap or bandanna will help keep dust or fibers out of your hair.

A paper respirator (sometimes called a "surgical mask") is important for keeping dust and fibers out of your nose and lungs. For eye protection, wear a pair of plastic goggles. When you're wearing both a respirator and goggles, the goggles may steam up as you work. To minimize this, try rubbing the inside of lenses with saliva, rinsing, and shaking them dry.

Tools. Buy an inexpensive work pouch to wear around your waist. By keeping your tools handy, it will save you many steps. For measuring, have a recoiling metal tape measure. And you'll find a retractable utility knife essential for cutting batts, blankets, or vapor-barrier material.

Either carry several new replacement knife blades in your pouch, or keep a sharpening stone handy. Sharp blades make a big difference, especially when you're cutting insulation that has a vapor barrier attached. If you use replacement blades, you'll also need a screwdriver for opening the knife. If a sharpening stone is your choice, keep it well oiled. Oil keeps the asphalt in vapor barriers from gumming up the knife blade. (Or you can draw the blade through the edge of a stud or joist to remove minor deposits from it.)

Where you must install a vapor barrier, you'll need a lightweight, squeeze-type stapler for fastening the barrier to wall studs, joists, or rafters. Be sure to get a stapler if you'll be fastening foil-faced or kraft-faced insulation to wall studs. Load it with ⅜-inch staples; you'll have to refill

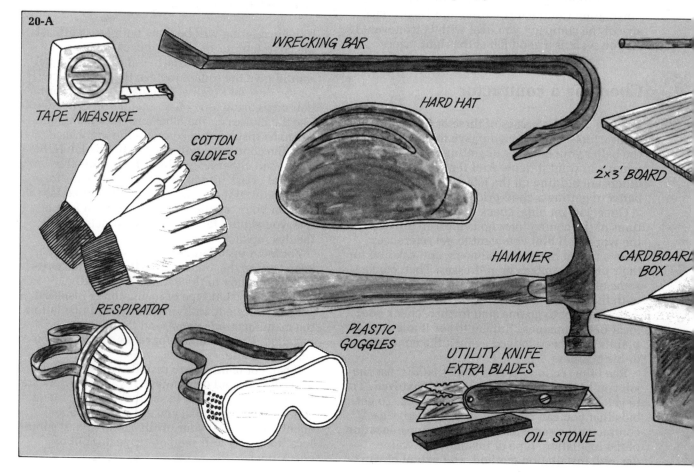

20-A

TAPE MEASURE

WRECKING BAR

HARD HAT

2"x3" BOARD

COTTON GLOVES

HAMMER

CARDBOARD BOX

RESPIRATOR

PLASTIC GOGGLES

UTILITY KNIFE EXTRA BLADES

OIL STONE

frequently, so keep the staple box in your work pouch.

For working in dark areas—in attics or under floors—you'll need at least one portable light with a 50-foot cord (a mechanic's "trouble light" works best). Twine is handy for bundling precut batts together for easy carrying.

Of course, for getting into an attic or reaching the tops of walls, you may need a stepladder. A stick or pole for pushing batts into out-of-the-way places also comes in handy.

About attics. Insulating attics calls for some special equipment. If you must kneel while working, wearing knee pads or wrapping your knees with elastic bandages will help keep them from getting bruised.

Because the ceiling between the joists won't support you, you'll need something more than ceiling joists to kneel and walk on. A couple of 2-foot by 3-foot boards laid over the joists serve well as temporary flooring.

It's a good idea to wear shoes or boots with nonslip soles. Also consider wearing a hard hat to protect your head from protruding nails. If you have to pry up any flooring, you'll need a hammer and a wrecking bar.

For blowing cellulose into an attic, you will need a blowing machine. You may be able to borrow one of these from the dealer who sells the insulation. If your dealer doesn't have one to lend, you should weigh the cost of renting one from a tool supply company against the time this method can save you. It may be nearly as easy to insulate with batts or blankets.

If you are going to blow in cellulose, you'll also need an assistant to load the machine. Outfit your helper with gloves and a mask. For some jobs, a 10-foot pole (bamboo is ideal) taped to the last 10 feet of flexible hose helps with blowing into hard-to-reach spots.

For hand pouring loose-fill fibers into an attic, obtain a large cardboard box with a lid; you'll need it for shaking the material to its full volume. A board or a bamboo garden rake makes spreading the material much easier.

About carpentry. If you will need to open up any walls, fur out wall studs, or do work of this nature, you'll need miscellaneous carpentry tools. The particular tools necessary are given in the how-to-do-it sections where techniques for installing various types of insulation in specific areas of a house are discussed.

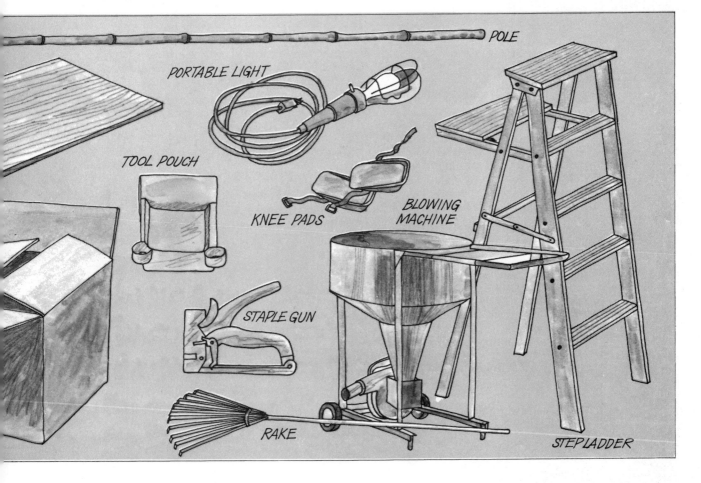

Ventilation... let your house breathe

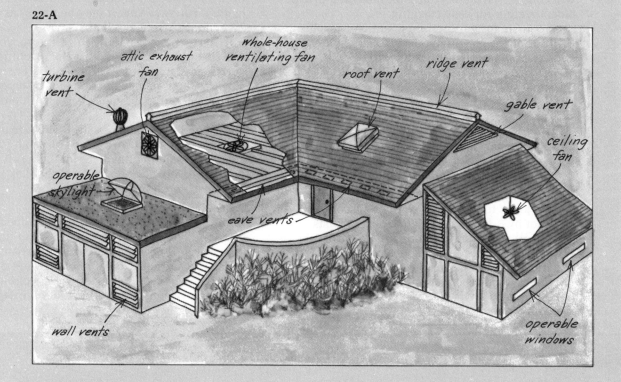

The more airtight you make your house by weatherstripping and caulking it, the greater your need for proper ventilation. If your house can't "breathe," unwanted heat, moisture, smoke, fumes, and vapors can build up inside. Vents and ventilating fans direct fresh air throughout a house, dispelling unwanted air and pollutants. Drawing **22-A** shows a variety of vents and fans.

Attic ventilation

Proper attic ventilation is most important for minimizing heat buildup and ridding a house of unwanted moisture (see "Why your house needs vapor barriers," page 24). On a hot day, trapped air in a poorly ventilated attic may push the temperature above 150°. Conducted and radiated to the living space below, this heat lingers long after the outside temperature drops. In winter, moisture condenses in an improperly vented attic, ruining insulation, staining ceilings, and damaging house structure.

Vents and fans both alleviate these problems by circulating outside air through an attic.

Attic vents are not mechanical. Most are simply openings covered by screens or grillwork. They are positioned to create an air flow pattern, as shown in drawing **22-B**. In this example, cooler

22-B

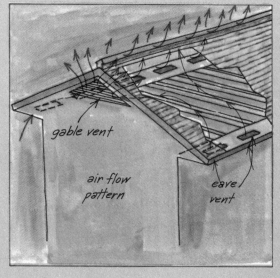

outside air enters through soffit or eave vents near the attic floor, forcing warmer, moisture-laden air through vents in the gables, roof, or ridge.

The proper type, number, and placement of vents depends on many factors, such as wind direction, hot spots caused by the sun, and roof lines that interrupt air flow.

As a rule of thumb, provide 1 square foot of free vent opening for each 150 square feet of attic floor area. If the attic has vapor barriers (see page 24), 1 square foot of free vent opening for each 300 square feet of attic floor space may be enough.

"Free vent opening" means an opening without wire or grillwork. You should subtract the area taken up by wire or grillwork. A vent with ⅛-inch or ¼-inch wire mesh should be 1¼ times as large as basic size. A vent covered by ¹/₁₆-inch insect screening (or ¼-inch mesh and a louver) should be twice as large. And a vent covered by ¹/₁₆-inch mesh and a louver should be three times as large.

A ventilation contractor can help you determine whether your house needs additional vents and—if it does—their sizes and locations. Find contractors in the Yellow Pages under "Contractors, Heating and Ventilating." Talk to several. If you plan to have a contractor do the work, get bids from at least two.

Attic fans. Vent openings don't always provide proper ventilation. An attic fan may be needed to push hot air through vents. One type often used is an *exhaust fan.* Installed in the roof or gable, it may be above an existing vent or in place of a vent. These fans, sometimes called "powered attic space ventilators," generally operate on thermostat controls that save energy.

Another type of fan is called a "whole-house ventilating fan." Installed in the attic floor, it cools the entire house. It does this by drawing cool air through open windows and vents in rooms below, pulling that air up into the attic, and forcing it out through the attic vents. Though rather expensive, these large fans are more efficient than roof or gable fans. In moderate climates, they are sometimes used in place of air conditioners.

In homes cooled by air conditioning, both exhaust fans and whole-house fans save more energy than they use by reducing the load on air conditioners.

Ventilating no-attic houses

In houses without attics, rising heat can get trapped near the ceiling. To release the rising hot air, you can place vents, windows, or skylights along the tops of walls or in the ceilings. If you do this, install low vents or windows as well to bring in cool air. If possible, position these openings to take advantage of prevailing breezes.

For hard-to-reach spots, choose windows that open with a long cord or hooked pole. In cold climates, consider double-glazed, operable windows to help retain winter heat (see page 78).

To get some cooling air movement, you can install either an old-fashioned ceiling fan or a standard recessed ceiling fan that vents out through the roof.

Venting household pollutants

Several special areas in a house require their own fans and vents. A kitchen, for example, needs a fan to exhaust smoke, heat, water vapor, and cooking odors. Your kitchen probably has an exhaust fan, located in the range hood or in the ceiling over the cooking area. If this fan doesn't move air satisfactorily, consider installing a second one in the wall or ceiling, vented outside through a duct. Be sure to keep kitchen fans clean.

Other places to install ventilating fans include bathrooms, laundry rooms, and other areas that collect large amounts of moisture or vapors. Always vent these to the outside through metal ducts.

Choosing a fan

When choosing a fan, consider two important features: capacity to move air and noise level.

Movement of air through a fan is measured in cubic feet per minute, abbreviated CFM. The loudness is measured in *sones*, an internationally recognized unit of loudness. The lower the number of sones, the quieter the fan. A 3-sone fan is twice as quiet as a 6-sone fan. One sone is roughly equal to the sound of a modern refrigerator operating in a quiet kitchen.

Investigate before you buy. Most fans are tested by the Home Ventilating Institute and given both CFM and sone ratings; these should be displayed on the fan or its container. Compare ratings on several different models before you make a choice.

A ventilation contractor or fan dealer can help you choose the right size fan for your needs. Detailed information on selecting and installing all types of ventilating fans is also available from the Home Ventilating Institute. Send a self-addressed, stamped envelope along with your inquiry to: Home Ventilating Institute, 230 North Michigan Avenue, Chicago, IL 60601.

Why your house needs vapor barriers

Showers, cooking, washing, and just breathing can put a surprising amount of water into the air in a typical home—5 to 10 pounds a day. If you wash and dry clothes, you may be adding another 30 pounds.

Because heat always moves to a colder location, in winter this warm, moist interior air passes through walls, roofs, and floors. As it goes through, the moisture condenses on the cold inner faces of the exterior surfaces. And there it accumulates. Eventually it blisters outside paint, forms stains inside, and damages the house's structure. And it saturates insulation, making the insulation practically useless.

A moisture barrier keeps moisture from passing through insulation and collecting inside walls, ceilings, and floors. It repels the moist air before it gets to the cold part of the wall (see drawing **24-A**).

24-A

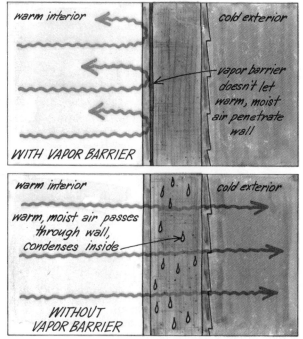

Ventilation plays a key role in determining whether or not a vapor barrier is necessary. If a crawl space or attic is properly ventilated, a vapor barrier may not be necessary (particularly in a dry climate). The warm, moist interior air takes the easiest route to the cold outside—in this case, through vents or fans.

Proper ventilation means at least 1 square foot of free vent opening for each 150 square feet of floor area. Vapor barriers aren't a substitute for ventilation, though. An area with a vapor barrier should still have 1 square foot of free vent for each 300 square feet of floor area (these figures are for attics). For more about ventilation, see page 22. If you live in an extreme climate or if you are in doubt about the need for vapor barriers, check with your building inspector.

Types of vapor barriers. You can get blanket insulations with attached foil or kraft paper vapor barriers. If you are insulating a space where you can use this type of insulation, it is usually the easiest solution.

If you are insulating with loose fill or with blanket insulation that doesn't have an attached barrier, you can install a separate barrier. For this, you can use either 2-mil (or thicker) polyethylene, laminated asphalt-covered building paper, or foil-backed gypsum wallboard (the latter is for walls and ceilings only).

Where vapor barriers go. You must be very careful to place the vapor barrier on the correct side of the insulation. If you put it on the wrong side, you will compound moisture problems instead of preventing them.

Put the vapor barrier toward the warm-in-winter side of the insulation.

When insulating between roof rafters, face the vapor barrier down. In an unfinished attic, between ceiling joists, the barrier should face the living space below. It belongs on the inner (room) side of exterior walls. And, when you insulate under floors, face the vapor barrier toward the room above.

A house is like a giant heat pump, pulling moisture out of the ground. Because of this, you should cover the ground of any crawl space

24-B

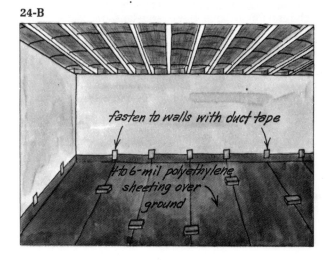

under your house with 4 to 6-mil polyethylene sheeting, overlapping it at the joints and anchoring it down with rocks or bricks to seal out excessive ground moisture (see drawing **24-B**).

Vapor-barrier alternatives. In some situations, such as walls covered on both sides, you don't have access for putting in a vapor barrier. Installing small louvered vents sometimes answers the moisture problem (check building codes for requirements in your area).

More commonly, the answer is to paint interior wall surfaces with 2 coats of a paint that has a "perm rating" of less than 1. Brush it in well. If you can't find such a paint, use two coats of high-gloss enamel or a varnish-based pigmented wall sealer, followed by a coat of alkyd paint.

Another alternative is to hire a contractor to foam in insulation. Foam insulations create their own vapor barriers (see page 14).

How to install a vapor barrier. Between floor joists in an attic, or under a floor that is over an unheated, poorly-ventilated space, lay in and staple strips of polyethylene as shown in drawing **25-A**. Cut the strips about 4 inches wider than the distance between joists, so you can fold the edges up before stapling. The insulation then goes on top of the vapor barrier. If you accidentally tear the sheeting, repair it with heating duct tape or electrician's tape.

For walls or attic rafters, you can stretch a continuous sheet of polyethylene across the faces of

staple strips of polyethylene between joists

studs or rafters and then staple it, as shown in drawing **25-B**. Staple across the top first. Be sure to keep the sheet taut and flat so the wallboard applied over it will lay flat. Hammer down any raised staples. Cut away the barrier around openings and electrical fixtures and outlets, and tape it around them.

When installing blanket insulation with attached vapor barriers, cut each piece a little long so the barrier will be complete from end to end. Staple flanges carefully along stud or joist sides.

25-B

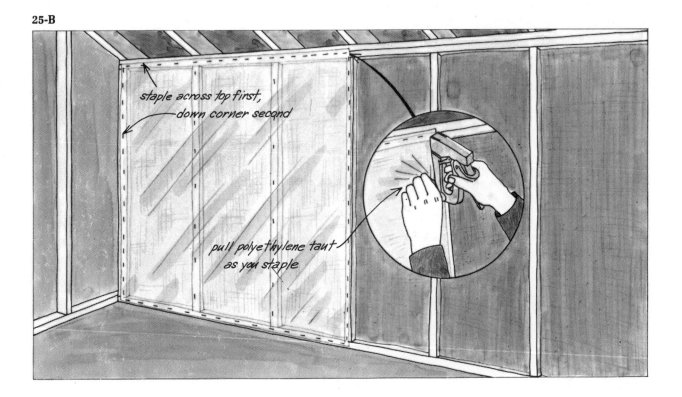

staple across top first, down corner second

pull polyethylene taut as you staple

Insulating an unfinished attic

If your attic is not serving as living space and you don't intend to use it for living space in the future, consider insulating it (or adding to insufficient insulation) as a top priority. Not only is an unfinished attic one of the easiest areas to insulate, it yields the greatest energy savings relative to cost.

Unless your unfinished attic has a floor, insulating end walls, knee walls, dormers, and similar areas is unnecessary. You need only insulate between the attic floor joists (see drawing **26-A**). The insulation slows the transfer of heat through the attic floor. An attic insulated this way is usually cold in winter and hot in summer, but these conditions shouldn't matter if it's not used as living space. If these conditions are unacceptable, insulate the attic rafters instead.

You can insulate an unfinished attic with either blankets (or batts) or loose-fill insulation. The discussion of the way to do the job with blankets (or batts) begins on page 28. For information on insulating an unfinished attic with loose fill, see page 30.

Making a choice

Should you install blanket-type insulation or loose fill? Each has its merits. Though installing blankets is usually easier than hand pouring loose-fill insulation, installing loose-fill cellulose is probably easiest of all if you can borrow a blowing machine from your insulation dealer. Though some preparation is needed around vents, lights, and so forth, cutting and fitting are not required. And carting bundles of insulation up into the attic is unnecessary—the machine moves the insulating material for you.

But the main problem with any type of loose-fill insulation is that it doesn't automatically provide a vapor barrier. If you're in a dry climate and your attic is properly ventilated (see page 22), you can probably get by without one. But under some circumstances, a vapor barrier is a must (see page 24). Where that is the case, in order to use loose-fill insulation you must provide a separate vapor barrier. It is often easier to

26-A

insulate an unfinished attic between the floor joists

use batts or blankets with attached foil or kraft paper vapor barriers.

Another consideration: loose-fill cellulose has a higher R-value than do blanket-type rock wool and fiberglass. Because of this, a minimum depth of 5 inches of cellulose will achieve R-19. It takes a minimum depth of 6½ inches of blown rock wool or a minimum of 8¾ inches of blown fiberglass insulation to produce the same R-value. The shallower depth will make a difference if joists are small and you want to add flooring without compressing the insulation or furring up the joists.

Loose-fill insulation can be installed directly over existing insulation. Batts and blankets can too—just use the type without a vapor barrier, or slash lots of cuts into the vapor-barrier backing.

What if there's flooring?

If your attic has a floor, you have two ways to insulate it yourself without taking up the entire attic floor: insulate the rafters and walls or blow in cellulose between floor joists. A third option is to have a contractor do the job.

Insulating rafters and walls is very similar to insulating a finished attic, as discussed on page 34. The main difference is that you don't have to

27-A

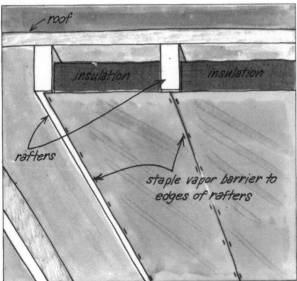

roof

insulation   insulation

rafters

staple vapor barrier to edges of rafters

contend with a ceiling surface—you can staple batts or blankets directly to the edges of rafters, as shown in drawing **27-A**.

To blow in cellulose, you'll have to pry up enough floor boards to gain complete access to the spaces between joists. Wood blocking is probably nailed between each pair of joists at one

27-B

to prepare for blowing insulation into attic, pry up flooring on both sides of blocking

drill ¼"-diameter holes at 1' intervals to prevent moisture build up

blocking

joist

or two places along their span. You'll need to pry up floorboards on both sides of the blocking (see drawing **27-B**). For more about this see "Installing loose-fill insulation," page 30.

Because you can't install a vapor barrier when insulating this way, the attic must have proper ventilation (see page 22). In addition to regular ventilation, you can prevent moisture buildup beneath flooring by drilling ¼-inch-diameter holes through the attic flooring at 1-foot intervals.

Getting started

Before installing insulation, prepare yourself by going through the following steps.

1) Read the information on vapor barriers, page 22, and on ventilation, page 24.

2) Take some basic measurements in your attic to determine the right insulation. Combining measurements with the vapor-barrier information, decide whether to use blankets, batts, or loose fill. If blankets or batts are your choice, figure the proper width and depth to buy (pages 16 and 17).

3) Calculate the proper quantities of insulation to purchase, using the information on page 13.

4) Read the information on tools and supplies, page 20, and gather the things you need. Take note of anything you don't have. You can pick up any supplies that you're fairly sure you'll need when you buy the insulation.

5) Call a few building supply stores and lumberyards to see who sells the right material at the best price. Prices vary; insulation is often a "lead item" for sales. Find out if the dealer will deliver or if you'll have to pick the insulation up. Then buy it and, one way or the other, get it home.

installing blankets and batts

Put on some old, loose-fitting clothes and set up your stepladder under the attic access hole. If the hole is over a closet, remove any clothing stored there to protect it from insulation dust and attic filth.

If you're working with batts, boost a couple of bundles up into the attic. You can do any necessary cutting on a board in the attic. If blankets are your choice, decide whether it's easiest to cut them in the attic or somewhere below. (Generally it works best to cut them in the attic.)

Put all the necessary tools in your tool pouch and take the portable light and your kneeling boards up into the attic. Hang the light where it's out of the way but will shed light on your work area. Span the joists with the kneeling boards; **walk only on joists or the boards**. Plan to work from the outer edges of the attic toward the center.

28-A

vapor barrier faces down

Start in the corner furthest from the access hole. If your blankets or batts have an attached vapor barrier, lay the first one between the joists with the barrier facing down (see drawing **28-A**). If you're working with a separate vapor barrier, fasten it to sides of joists and then lay batts or blankets on top of it (see drawing **25-A**, page 25).

Be sure the first batt or blanket extends over the top plate of the wall below. But don't cover eave vents or block their flow of air. Their ventilation is necessary. (See drawing **28-B**.)

Don't cover anything that can produce heat, or

28-B

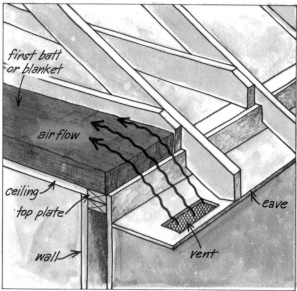

first batt or blanket

air flow

ceiling

top plate

wall

eave

vent

you'll create a fire hazard. This includes electric fans and recessed light fixtures (drawing **28-C**). (Recessed light fixtures are those with an enclosed metal box or canister holding a light bulb above the ceiling line.) Also peel back the flammable vapor barrier 3 inches from chimneys, stovepipes, and flues.

28-C

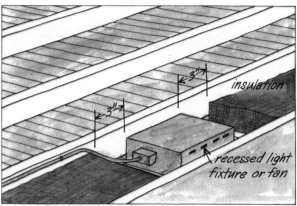

insulation

recessed light fixture or fan

Below-ceiling light fixture boxes and other electrical boxes are safe to cover, but don't hide electrical junction boxes where access may be necessary. And don't attempt to bend or pull electrical wiring out of the way—slip your batts underneath where possible.

If you are installing blankets, insulate the long runs first to minimize cutting and waste. Unroll the first blanket until it encounters a wood block or the center of the joist space; then cut it off, using the method described at right. For precut batts, butt a second batt up to the first, overlapping vapor barriers slightly. Work from both ends of the joist spaces to the center of the attic, where it's easier to measure, cut, and fit. When insulation encounters a block or crossbrace, cut it to fit snugly (see drawing **29-A**).

29-A

cut to fit around crossbracing

Repair any tears in the vapor barrier with heating-duct tape or electrician's tape. (Where building codes don't require a vapor barrier in a properly ventilated attic, you don't need to worry about overlapping the barrier or repairing tears.)

Then go on to the next row between joists, and work your way across the attic doing this. Air space between the insulation and ceiling increases the insulation's effectiveness (see drawing **29-B**). If blankets or batts rise above the top

29-B

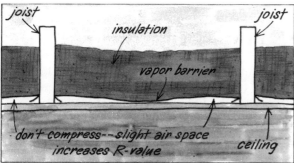

joist — insulation — joist

vapor barrier

don't compress—slight air space increases R-value — ceiling

edges of the joists, don't compress them by laying temporary flooring on top. Compression decreases their R-value.

Install the insulation carefully, fitting it tightly around all edges. Using loose insulation pulled from scrap pieces, stuff any places where the blankets or batts don't fit snugly.

If you need two layers of blankets or batts to achieve a high R-value, the second layer should not have a vapor barrier, or its vapor barrier

29-C

no vapor barrier

second layer goes perpendicular to first layer

first layer of insulation

to support flooring, you must add blocks to joists

should be slashed with lots of cuts. It should be laid at right angles to the first layer (see drawing **29-C**).

Cutting

Cut insulation on a scrap sheet of plywood or an unfinished subfloor. With a little practice and a confident stroke, you can cut straight across a batt without the aid of a straightedge. It's usually easiest to cut with the vapor barrier facing up.

To cut off a piece, first stretch your measuring tape along the length of the insulation's far edge for a measurement. As shown in drawing **29-D**, insert the blade of your utility knife just shy of the far stapling flange and cut toward you, right through the flange near you. Spread the fibers

29-D

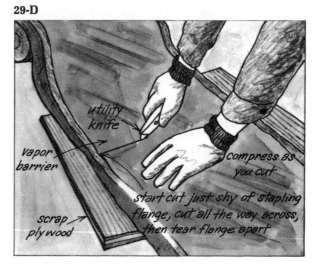

utility knife

vapor barrier

compress as you cut

start cut just shy of stapling flange, cut all the way across, then tear flange apart

scrap plywood

apart with your free hand as you cut with short strokes through the insulation into the wood. Then tear the far stapling flange apart with your fingers. (If you are cutting unfaced insulation, you should, of course, ignore the directions that deal with the flanges and barrier.)

To cut a piece lengthwise, guide the knife along a scrap of lumber, such as a 1 by 4 (see drawing **30-A**). Be sure to allow an extra inch in

30-A

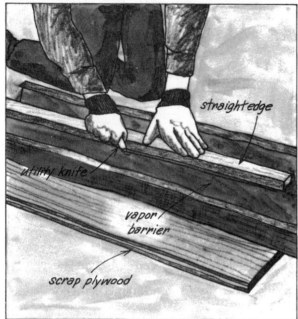

width so the insulation will fit snugly and, if it has a vapor barrier, so it can be stapled along both edges.

Finishing touches

Before you leave the attic, staple insulation onto the top of the attic access cover, vapor barrier down, as shown in drawing **30-B**. Then take a last look around to make sure the job is complete and you have not covered any vents, recessed lights, or fans. Gather up your tools and equipment and climb down. Throw your work clothes into the washing machine by themselves, then take a cool shower to wash away the insulation's itchy aftereffects. (Warm water opens your pores, letting fibers settle into your skin a bit more.)

30-B

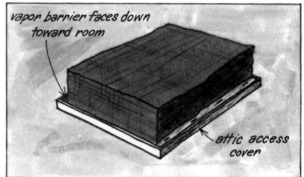

installing loose-fill insulation

Before you begin any work, read the introductory information under "Insulating an unfinished attic," starting on page 26. Pay special attention to the preliminary steps listed under "Getting started." *Be sure to install a vapor barrier if one is needed.*

Preparing the attic

Set up your stepladder under the attic access cover. Remove any clothing or other items stored beneath it so they won't get soiled. Climb up inside to see how much flooring—if any—is nailed to the tops of the floor joists. **Do not step on the ceiling between the joists—it won't support you.** If there is no flooring, you can skip the next step.

If the attic has a floor, you will need to take up some of the flooring boards so you can blow insulation into the joist spaces. This means prying up enough boards to give you a clear view through the spaces between each set of joists, so that you can check for any blocks.

You'll need a wrecking bar for prying up the

30-C

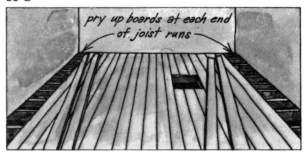

flooring, a light, and a hand-size mirror to make the job of looking easier. Pry up one or two boards at each end of the joist runs (drawing **30-C**). Shine the light in one end and look into the other end (reflecting your view with the mirror eliminates the need to stick your head down inside).

Take up the flooring directly on each side of any block and repeat the checking process until you have direct access to all of the spaces.

The next step is to count the number of eave vents your attic has. Eave vents are screened openings located at the ceiling line, along lower edges of the roof where the rafters meet the floor joists. Because they are a necessary part of your house's ventilation system, you must be sure they don't get blocked with insulation.

The best way to insure this is to make baffles of wood or some other material and fasten them securely between joists at the soffit on a 45-degree angle, as shown in drawing **31-A**.

31-B

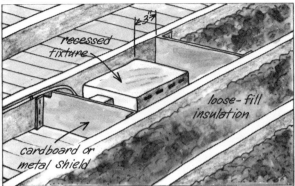

recessed fixture

loose-fill insulation

cardboard or metal shield

31-A

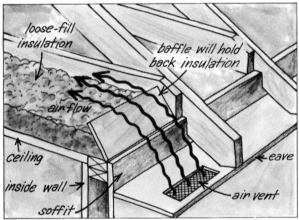

loose-fill insulation

baffle will hold back insulation

air flow

ceiling

inside wall

soffit

eave

air vent

An alternative method is to make a shield from a piece of 6-inch blanket insulation, about 2 feet long and the same width as the joist spacing. Place the batt between the joists, just shy of the vent (see drawing **28-B**). If the insulation has a vapor barrier, face it down.

Next, locate any electrical fixtures that must be sheltered from the loose-fill insulation—specifically electric fans and recessed light fixtures (the metal canister type that hold the light bulb above the ceiling line). The National Electrical Code requires that insulation be kept back 3 inches from heat-producing electrical equipment.

You can devise shields by fastening cardboard or sheet metal to joists with staples or nails (see drawing **31-B**), or by using spare pieces of 6-inch insulation batts or blankets. If the batts or blankets have a vapor barrier, tear it off (the vapor barrier is flammable). To further shield a fan or recessed light, put a temporary roof over it—one made of batting or cardboard. *Be sure to remove the "roof" after insulating.*

Next, separate chimneys and flues from loose-fill areas, using 6-inch blankets or batts. Tear off the vapor barrier if there is one. Leave a 3-inch space between the insulation and the chimney or flue.

Look for and repair any defective wiring. For complicated repairs, call an electrician. If a wire has a worn or cracked sheath of insulation, shut off the power and wrap the defective portion with electrician's tape.

Loose-fill insulation sifts through all cracks, especially when it is blown in. Tear pieces from batts and stuff them in any gaps and cracks. Cover each nonrecessed light fixture and common electrical box with a piece of batting to keep fibers or pellets from sifting down through the cracks around it. Staple batting to the top side of the attic access cover, vapor barrier side down, as shown in drawing **30-B** (facing page).

How to blow in insulation

Cellulose is the only insulation a homeowner can easily blow into an attic by machine. If you cannot borrow a blowing machine from your insulation dealer, you'll either have to rent one from a tool-rental agency or use a different method.

Blowing in insulation is much faster than spreading loose fill by hand; it also allows you to control the R-value more accurately. On the negative side, the work is dusty and you'll need a helper to load the blowing machine.

When you're ready to start work, wheel the machine as close to the access hole as you can. Park it on top of a plywood panel, tarpaulin, or old blanket. Stack the bags of cellulose nearby.

Before connecting the 3-inch hose to the blower, slide open the trap underneath the hopper. Reach up inside, as shown in drawing

check in trap under hopper for fibers

32-A, to remove any fibers left by a previous do-it-yourselfer.

Plug in the electrical cord. Turn the blower switch to "slow." Hold your hand over the vent to make sure the blower works. Then turn off the blower switch and connect the hose to the machine.

On most machines, the bags rest one at a time on a metal handle; pull it out and lock it into place. Using your utility knife, cut several bags of insulation midway between top and bottom. Pull the halves completely apart to make loading easier. Rest one half bag on the handle, open end facing the hopper (see drawing **32-B**).

32-B

rest open bag on handle

You and your helper should now put on your protective masks (see page 20). If you'll be the one working the hose in the attic, you should also put on a hat and goggles. Have your helper stay at the machine while you take the end of the hose and a portable light up into the attic.

Step only on the floor joists, flooring, or temporary platform boards. Move to the furthest point from the access hole and plan to work back toward the hole. Call for your helper to turn on the machine and to start loading it. Don't worry—the hose won't jerk and writhe like a hungry python; it should be easy to manage.

While you blow insulation into the spaces between the joists, your helper feeds cellulose into the hopper. To keep a steady flow of material going through the machine, the helper should scoop up chunks from inside the bag and break them into smaller pieces by hand or by rubbing them against the bottom grate. The rotating choppers in the machine complete the "fluffing" process.

Fill the spaces between joists to the depth required for the R-value you wish the attic to have (see page 12). Be sure to blow the cellulose completely into joist spaces. You can generally control the direction and velocity of the moving insulation by using your hand as a baffle at the hose's end (drawing **32-C**).

32-C

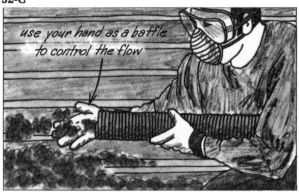

use your hand as a baffle to control the flow

For blowing into hard-to-reach spots, tape a pole to the last few feet of hose to stiffen it (see drawing **32-D**).

If the hose plugs, your helper should turn off the machine. Then reverse the hose, hooking up

32-D

tape pole to end of hose for hard-to-reach spots

your end to the machine. Turn it on to blow the hose free of blockage.

Before you blow insulation near the attic access hole, rest the cover on top of the hole as well as you can to limit the amount of insulation that will blow down through the hole.

How to pour loose fill

If you're unable to borrow or rent a cellulose-blowing machine, and if you prefer to hand pour loose-fill insulation, plan to spend quite a bit more time on the job. And remember that to achieve the same R-values, loose-fill insulation poured in place requires greater thicknesses than both cellulose and blanket-type insulations.

If you're using mineral-wool insulation, buy bags labeled "pouring wool," not "blowing wool." Blowing wool is packed too tightly to fluff by hand. Granular insulations—vermiculite and perlite—are made only for pouring.

After preparing the attic as previously outlined, boost several bags up into the attic. Take a bag to the corner furthest from the access hole. If you're using granular insulation, just cut the bag open and pour the contents between the joists where they meet the rafters (see drawing **33-A**).

33-A

simply pour granular insulation between joists

Work your way back toward the attic access hole.

For mineral-wool insulation, you will need a large cardboard box with a top. Wear the protective gear suggested for working with blanket-type insulations (see page 20).

Dump enough insulation into the box to fill it to about one-third of its volume. As shown in drawing **33-B**, put the lid on the box and shake vigorously until the insulation fluffs out to fill the box (three times its packed volume). Scatter

33-B

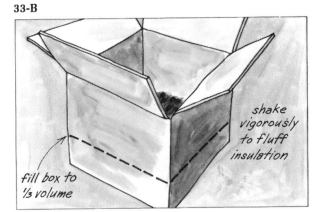

shake vigorously to fluff insulation

fill box to ⅓ volume

the fluffed fibers between the joists and repeat the process—again and again—until all joist spaces are filled to the proper depth (as deter-

33-C

use bamboo rake to spread loose fill evenly

mined by the information on page 12). Use a bamboo rake to spread loose fill evenly (drawing **33-C**).

Finishing touches

Before you descend from the attic, be sure to remove all temporary covers from over fans and recessed light fixtures. Do not remove the side baffles that hold back the loose fill from fixtures or vents.

If you've pried up any flooring, nail it back in place. Gather up all tools, equipment, and empty bags. Then climb down, replacing the attic access cover as you leave.

Vacuum up any insulation that has sifted down through the access hole. If you have used a blowing machine, vacuum around it and clean out any remaining fibers from the trap beneath the hopper before returning it to the dealer.

To wash away possible irritation from fiberglass or rock wool, take a cool shower. (Warm water opens your pores, letting fibers settle into your skin a bit more.) Throw your clothes into a washing machine by themselves.

Insulating a finished attic

Insulating an attic that has a floor, covered walls, and a finished ceiling is more complicated than insulating an unfinished attic. Why? Because the insulation must go between the living space and the outdoors.

In an unfinished attic, insulation simply lies between the joists that support the ceiling of the room below. But in a finished attic, you must put insulation in several places: up between roof rafters, in the space above collar beams, inside the room's end walls, to the outer side of knee walls, and between the attic floor joists on the outer side of the knee walls. Drawing **34-A** illustrates these locations.

Access is the main problem. Unless crawl holes already exist, you must cut knee walls open to insulate behind them. And to do a complete job, you must get insulation between rafters above the ceiling covering, open access holes for insulating between collar beams, and put insulation into end walls.

To do the job yourself, you must either force insulation blankets (or batts) up between the rafters, or you must blow in cellulose or hand pour granular loose-fill insulation from the ceiling's peak. If nails protrude down through the roof sheathing, as they often do, pushing or pulling blankets through the rafter spaces is nearly impossible. Loose fill is easier to install in these spaces, especially if you can borrow or rent a machine for blowing it in.

The main problem with loose fill, however, is that you may have to install a vapor barrier sepa-

rately (see information on vapor barriers, page 24). In most localities, loose fill is recommended for finished-attic insulation only where a vapor barrier already exists or is not necessary.

With these considerations in mind, check your attic to see if insulating it yourself is feasible. Remember that some insulation is better than none. If hiring a contractor is too expensive, yet you can't insulate all of the areas yourself, do what you can.

The easiest areas to insulate are usually the outer sides of knee walls and the spaces above collar beams. If the attic ceiling area is inaccessible and your house needs a new roof, consider reroofing and having a contractor insulate under the new roofing material. Or fasten insulation board to the inside surface of the attic ceiling as you would insulate a beamed ceiling (you can adapt the methods given on page 38). But whatever you do, don't put in insulation and neglect the vapor barrier where it is needed.

The following discussion offers techniques generally used for insulating a finished attic.

Gaining access

The first step is to get behind the short knee walls. If no entry holes exist, you'll have to cut your own. Drawing **35-A** shows how a typical knee wall is built. For each knee wall, you cut away (and save) a section of gypsum wallboard between two wall studs. The resulting

34-A

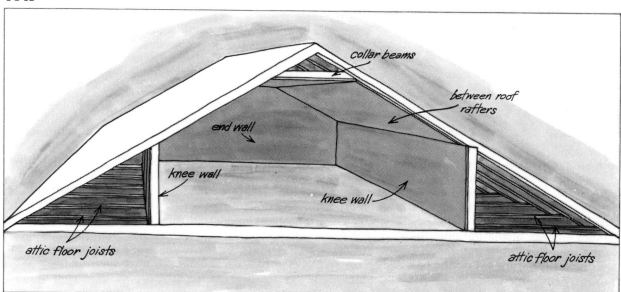

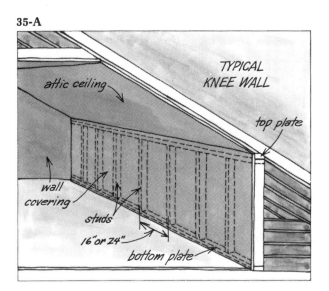

14½-inch-wide space is large enough for average-size people to squeeze through.

To cut out the panel, first mark its perimeter. You will need to locate the inner edges of two studs—preferably two that are near the center of the wall. An effective way to locate a stud initially is to use an inexpensive tool called a "stud finder." Its pointer fluctuates when passed over a nail (and nails almost always hold gypsum wallboard to studs). See drawing **35-B**.

35-B

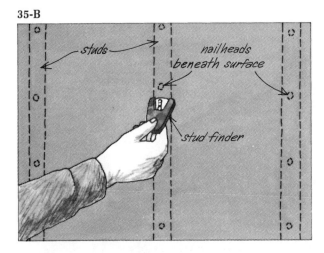

Once you find one stud, look for another 16 inches (or maybe 24 inches) to either side of it. Mark locations of two adjacent studs. These marks approximate the studs' centers.

Next, remove the base molding and measure up from the floor 1½ inches. Draw a horizontal line at this level between the two studs. This cutting line, 1½ inches up from the floor, allows the saw blade to cut above the wall's bottom plate (see drawing **35-C**).

Now draw another horizontal line between the

35-C

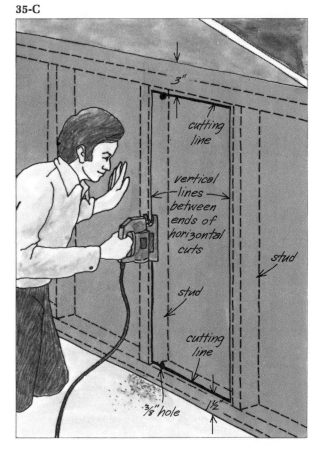

studs, either 3 inches down from the ceiling or, if the wall is taller than 4 feet, 4 feet up from the floor. A 4-foot-tall hole offers convenient access, yet keeps replacement of the panel easy.

Now you've marked your horizontal cutting lines. Drill a ⅜-inch hole to the inside of one of the horizontal lines and start a cut from it with a saber saw or keyhole saw (see drawing **35-C**). For faster cutting, you can finish the cut with a regular handsaw once the saw groove is long enough to insert the handsaw's larger blade.

Cut until you reach one of the studs, then reverse direction and cut to the other stud. Cut both horizontal lines this way.

Then draw two vertical lines connecting the ends of the horizontal cuts. Using the same methods, cut along these vertical lines. Lift out the rectangular panel and put it aside for later.

Now you can crawl behind the knee wall to survey the situation, taking a tape measure and a portable light. **Do not put your weight on the ceiling between the floor joists; it will not support you.** Crawl or step only on the joists or on temporary flooring boards. Look for and measure amounts of existing insulation and check any existing vapor barriers for rips.

Also try to sight up between the roof rafters, over the attic room, to see what your chances are of getting insulation between them. Nails protruding down through the roof sheathing will probably rule out use of blanket-type insulation for those spaces.

For installing loose fill between the rafters, or if you want to insulate the space above collar beams, you will need to gain access to the space at the attic ceiling's peak. In some houses, collar beams are only a foot or two long; in other houses they are so long you can crawl up over them. See drawing **36-A**.

36-A

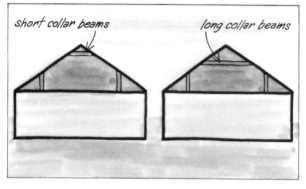

short collar beams   long collar beams

If your attic's collar beams are the longer type, you'll need to cut only one hole into the ceiling—one that you can crawl through. For attics with short collar beams (or none at all), you would have to cut a separate hole at the top end of each rafter cavity—holes big enough to stick a 3-inch hose into or to pour loose-fill granules into. (In most cases, one long hole is easier to patch than several small holes. See drawing **37-A**.)

Find rafters and cut access holes between collar beams, using the same methods as for cutting into knee walls.

For a thorough job, one other area should be insulated: the finished attic's end walls. For information on insulating a wall that is finished on both sides, see page 39.

Getting started

Once you've gained access to the various attic spaces, measure the square footage of insulation you need for each. Figure the type and amounts of insulation you'll need, referring to pages 13 and 16.

Next, read the section on tools and materials, page 20. Note the things you don't have and, if you are fairly sure you'll need them, plan to buy them when you get the insulation.

It pays to check insulation prices before buying. Call several building supply stores and lumberyards to see who has the material you want at

the best price. Find out if the dealer will deliver or if you must pick it up, then arrange to get it to your house—one way or the other.

Installing insulation

The areas to insulate first are those behind knee walls. For the spaces between the floor joists, use either loose-fill or blanket-type insulation. Because insulating these spaces involves the same steps as insulating floored, unfinished attics, see complete do-it-yourself information on page 26, under the heading "Insulating an unfinished attic."

Be sure you don't cover eave vents with insulation. And follow the precautions discussed in that section for recessed light fixtures (the type that hold the light bulb above the ceiling line), fans, flues, chimneys, and stovepipes. Insulation must be kept back 3 inches from heat-producing equipment.

If you can use batts to insulate the rafter spaces, push them up into place with a pole or stick before you insulate the knee walls. Use the type that has an attached vapor barrier; face the barrier down toward the room.

Otherwise, insulate the backside of the knee walls first. Install blankets with the vapor barrier facing in toward the room. Cut pieces long enough to reach from the bottom plate all the way up to the roof sheathing (see drawing **36-B**).

36-B

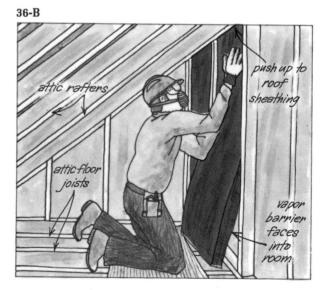

attic rafters

attic floor joists

push up to roof sheathing

vapor barrier faces into room

For information on cutting, see page 29. Because each batt goes up beyond the top of the knee wall, it can serve as a backstop for loose-fill insulation blown or poured into the rafter spaces from above. The batts keep the blown-in loose fill insulation from sliding off the sloping ceiling.

Staple the top ends of the blankets to the roof sheathing. The methods used for fastening the batts between knee-wall studs are the same as those for holding batts under floors. You have several choices, discussed in depth on page 43.

To insulate rafter spaces with loose fill, pour granular insulation or blow in cellulose from the access holes at the peak of the ceiling, as shown in drawing **37-A**. Install as much as you can. As

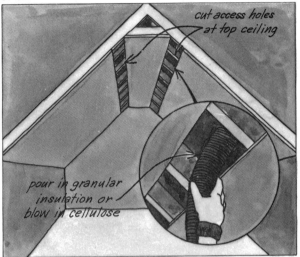

you work, check occasionally behind knee walls to be sure it isn't flowing out at the other end of the rafters.

Insulate the area over collar beams with either loose fill or blankets, using the same methods as those for insulating the attic floor outside the knee walls. When installing insulation with an attached vapor barrier, face the barrier down.

Insulation should fit tightly in all areas. Stuff any cracks or gaps with pieces pulled from scraps of blanket insulation.

Closing the access holes

Before closing access holes, be sure to take all tools and excess materials out of the areas you are sealing up. If you have used loose fill, remove any temporary covers you've placed over recessed lights or fans.

Start with the knee walls. Here's how to seal up each hole:

Fasten 2 by 4s around the inner perimeter of the opening to give you something to nail the wallboard to (see drawing **37-B**). These should be flush with the other studs, set back from the face of the wall by the thickness of the wallboard (usually ½ inch).

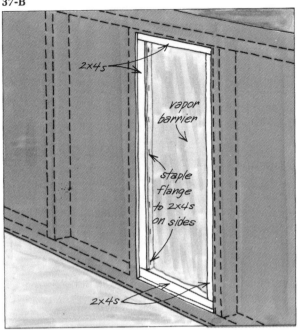

Staple a short piece of insulation between the rafters over the opening, from the top plate up to the roof sheathing. Then staple a piece of insulation in the opening, as shown in drawing **37-B**.

Replace the panel in the wall. Nail around the perimeter with gypsum-wallboard nails. Dimple the wallboard's face slightly with the last hammer stroke on each nailhead.

To hide the nails and cracks, first apply a smooth "bedding" coat of wallboard taping compound over all joints and nail dents, using a 6-inch drywall knife (drawing **37-C**). The molding will hide the lower horizontal joint. Before the compound dries, smooth drywall tape over the joints, applying a thin layer of compound over the tape. When this dries, apply a second coat of compound to joints and nailheads, feathering the edges. After the second coat dries, apply a last coat of compound with a 10-inch drywall knife, feathering the edges. Lightly sand the surfaces smooth, using 80-grit sandpaper, then paint to match the wall. Replace molding.

To patch ceiling openings, use the same methods.

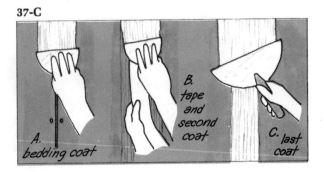

How to insulate a beamed ceiling

Because there is no space for standard insulation, cathedral-style or exposed-beam ceilings require special insulating methods. Drawings **38-A** and **38-B** show two methods you can use. Neither offers very high R-values.

As drawing **38-A** shows, you can place rigid insulation of varying R-values (usually R-9) above the wood decking, but this necessitates reroofing. Unless you are experienced at carpentry, leave this to a contractor. As an alternative in some localities, a contractor can foam urethane insulation onto the roof's surface—a technique that is akin to reroofing.

Drawing **38-B** shows a method you can do yourself. You place 2-inch-thick, vinyl-faced pressed-fiberglass panels up between the beams and fasten them on the underside of the roof decking. Several styles of textured vinyl faces are available on the panels. Because of their extremely light weights, these R-9 panels are easy to handle. They are available in sizes up to 4 feet by 16 feet.

Installation is simple. Using a kitchen knife, cut the panels to fit snugly between beams. Press them into place (see the photo below) and nail quarter-round molding around the perimeter of each one to hold it there.

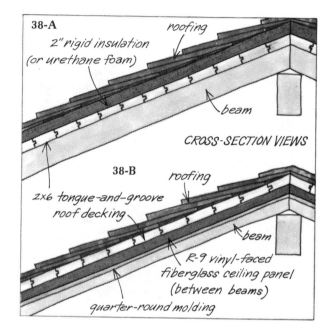

38-A
roofing
2" rigid insulation (or urethane foam)
beam

CROSS-SECTION VIEWS

38-B
roofing
2x6 tongue-and-groove roof decking
beam
R-9 vinyl-faced fiberglass ceiling panel (between beams)
quarter-round molding

Lightweight fiberglass panels *fit between beams; quarter-round molding secures them in place.*

How to insulate walls

Because walls make up such a large portion of the exterior of a house, they are second only to the roof in terms of losing or gaining heat. If you can insulate walls with blankets or batts during the construction of a house, you're well ahead of the game. But insulating walls after both inner and outer surfaces are attached is difficult—a job best left to insulation contractors who can foam insulation into them.

This section discusses what is required to insulate finished walls and why you should leave the work to professionals. It then gives you step-by-step techniques for insulating unfinished walls in your house.

insulating finished walls

If you live in an older house, the chances are good that the exterior walls have insufficient insulation or none at all. To put insulation into walls that are surfaced on both sides, you must gain access into the stud cavities, from inside or outside the house. Because this job is difficult and messy, you should weigh the possible savings (up to 20 percent in fuel bills) against the headaches and cost of having the job done.

To insulate a wall from inside the house, the wall's surface—wallboard or paneling—may be removed so insulation batts or blankets can be installed between the wall studs as though it were an unfinished wall (see page 40 for techniques). Of course, doing this involves major work, akin to gutting your house. It is only practical if you are completely remodeling.

Instead of removing entire sections of wood-frame wall surfaces, contractors normally cut holes through either the interior or exterior surfaces and blow in or foam in insulation (drawing **39-A**).

39-A

foam is injected through holes in siding

CUT-AWAY VIEW

If the siding is clapboard or shingle, the holes are usually bored from the exterior. But if the exterior walls are of masonry, metal siding, or a material that is especially difficult to patch, the holes are cut into the interior walls.

Drilling the holes is not a minor job. A 1½-inch to 2-inch hole must be bored into each cavity between each pair of wall studs—both at the cavity's top end and below each fireblock or obstruction. For a two-story house, this must be repeated on the second story (see drawing **39-B**).

39-B

holes for foaming insulation into a typical two-story house

Most contractors charge less if you do this preparatory work for them. But the work is considerable: removing sections of siding or wall surfaces, drilling holes at the top of each cavity, dropping a plumb bob down through each hole to probe for obstructions, and drilling holes under obstructions. In addition to this, there's the chore of patching afterwards. Talk with your contractor before doing any of this.

Foamed-in insulation has high R-values per

inch of thickness, creates its own vapor barrier, and totally fills cavities. Remember that some foams are flammable or toxic when ignited; building codes must be strictly followed.

Blown-in fibers—and even cellulose—catch on nails, pipes, and other obstructions and only partially fill cavities (see drawing **40-A**). Some types settle with time. And they don't automatically provide a vapor barrier (see page 24). To create a vapor barrier when using them, you must generally paint interior wall surfaces with two coats of glossy oil-based enamel or a pigmented wall sealer, followed by one coat of alkyd.

Of course, the methods of insulating discussed here are for standard wood-frame walls. If you wish to insulate existing concrete or concrete-block basement walls, see page 46.

BLOWN WOOL R-11   FOAM R-19

insulating unfinished walls

A standard wall that doesn't have a surface on one side of the studs is usually the easiest place to insulate. This is because it is so accessible.

Where in a house are there such unfinished walls? Often between the main part of the house and an attached garage or utility room, in an attic, or in a basement. Or, if you are in the process of building a new home or remodeling an older one, all of the walls might be unfinished.

Unless you wish to use insulation as a sound barrier, the only unfinished walls you should insulate are the ones between heated living spaces and unheated spaces or the outdoors.

Mineral-wool batts or blankets are best for insulating unfinished walls. Precut 4-foot batts are easiest to handle when insulating a standard 8-foot wall, but blankets can be cut to length easily. It is generally best to choose the type with an attached vapor barrier.

The instructions that follow are for installing blankets and batts, but don't rule out the use of rigid-board insulation, discussed on page 18.

When installing insulation with an attached vapor barrier, the barrier should face the side of the wall that's warm in winter. Because of the flammability of vapor barriers, building codes require that they not be left exposed. So, where a vapor barrier faces the unfinished side of a wall, plan to cover it with gypsum wallboard or some other fire-retardant wall surface.

Where a vapor barrier is not required, you can use unfaced batts. Some types need not be fastened into place; they wedge tightly between studs.

Getting started

Here are some preliminaries to take care of before you begin insulating.

1) Read the information about vapor barriers, page 24, and ventilation, page 22.

2) Decide whether to buy batts or blankets (see page 16) and whether to get the 16-inch or the 24-inch width. Measure the walls you will insulate and compute the amounts you'll need, using the information on page 13.

3) Read about necessary tools and supplies, page 20, and gather the things you will need. If you are missing anything that you think you will need, plan to get it when you buy the insulation. (If much of your work will be out of reach, consider renting a folding aluminum scaffold.)

4) To buy insulation at the best price, call several building supply stores and lumberyards. Determine whether the dealer will deliver the material or if you will have to pick it up. Then buy it and, one way or another, get it home.

5) Climb into some old, loose-fitting clothes, put all the necessary tools in your tool pouch, and gather your materials near the place you will be working.

How to cut blanket-type insulation

You can cut batts or blankets to length on a subfloor (or a piece of plywood) or after they are partially placed in the wall. Most professional

installers cut 4-foot batts in the wall with one end pushed snugly against the wall's top plate or bottom plate and the other end draped across the fireblock, which is used to back the cut (see drawing **41-A**).

No matter how you do the cutting, size the pieces slightly long to insure a tight fit. With a little practice and a confident stroke, you can cut straight across a blanket or batt without the aid of a straightedge. It's usually easiest to cut with the vapor barrier facing up. Insert the blade of your utility knife shy of the far stapling flange and cut toward you, right through the flange near you (see drawing **29-D** on page 29). As you cut with short strokes, spread the fibers apart with your free hand, being sure the blade cuts all the way through. Tear apart the far stapling flange with your fingers. (If you are cutting unfaced insulation, there is no barrier to worry about.)

To cut lengthwise along a batt or blanket, guide the knife with a scrap of lumber, such as a 1 by 4, as shown in drawing **30-A** on page 30. Allow about an inch extra in width so the insulation will fit snugly and, if it has a vapor barrier, so it can be stapled along both edges.

Installing batts and blankets

Starting at one upper corner of a wall, push a batt between the studs. If the wall is a little too high for you to reach the top, nudge the batt into place with a stick or pole. Push several batts into place, then go back and fasten them.

An attached foil or kraft-paper vapor barrier (if there is one) should always face the side of the wall that's warm in winter. If it faces you during installation, you can fasten the batts using a stapling gun with ⅜-inch staples. Drive staples through the stapling flanges into the sides of the studs every 12 inches (drawing **41-B**). Tuck in the end at the fireblock as shown.

If the vapor barrier must face away from you during installation, fasten the batts in place with the methods discussed on page 43 for installing batts or blankets under floors.

When installing a batt that has been cut lengthwise, staple it along the vapor-barrier flange remaining along one edge. Then pull the cut edge of the vapor barrier across to the other stud, pushing away the insulation behind it, fold slightly, and staple through the barrier into the stud. See drawing **41-B**.

41-B

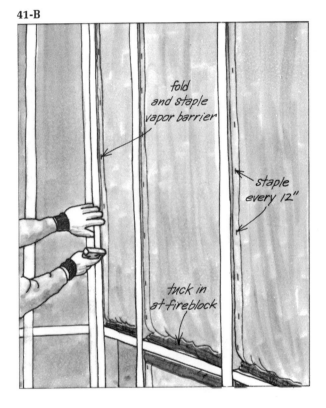

Peel back flammable vapor barriers 3 inches from flues, chimneys, electric fans, and other heat-producing equipment.

By placing insulation both behind and in front of water pipes, you can muffle gurgling water and protect the pipes from freezing in winter. Be sure insulation is between the pipes and the outside.

Stuff insulation into cracks and small spaces between rough framing and the jambs, heads, and sills of windows and doors. Also stuff spaces behind conduit, electrical outlet and switch boxes, and other obstructions. For this, use fibers pulled from scrap pieces of blankets or batts.

To clean up, throw your work clothes into a washing machine by themselves and take a cool shower. (Unlike warm water, cool water helps keep fibers from entering your pores.)

Insulating under floors

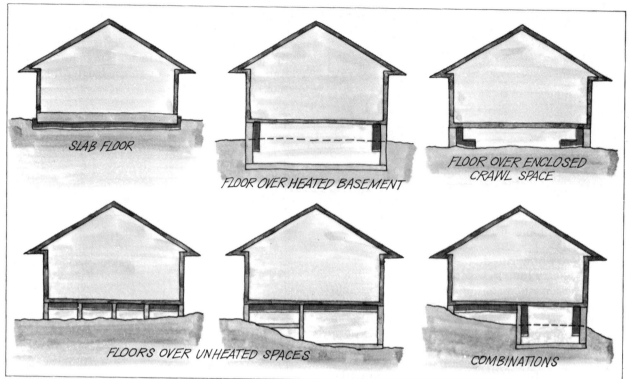

SLAB FLOOR

FLOOR OVER HEATED BASEMENT

FLOOR OVER ENCLOSED CRAWL SPACE

FLOORS OVER UNHEATED SPACES

COMBINATIONS

If heating bills are a real problem where you live, don't neglect insulating the areas under floors. Savings on summertime air conditioning are minimal, but you can save substantially on winter fuel costs. Heating a basement or crawl space requires as much energy as heating other rooms; an unheated basement or crawl space draws heat from the floor above, causing drafts and making the floor cold.

Houses can have any of several kinds of floors. Your method of insulating will depend upon how your house sits on the ground. Drawing **42-A** illustrates several situations and shows where insulation goes for each.

If your house has a concrete slab floor that sits directly on the ground, insulating—or even checking for existing insulation—is very difficult, if not impossible. Under most circumstances, rule out the possibility of insulating this type of area.

If your basement is heated and the basement walls project out of the ground, insulating may be advantageous. See page 46 for information on this. The walls of an enclosed crawl space should be insulated too; see page 45.

The following discussion covers insulating floors over unheated basements, garages, porches, or crawl spaces. Because these areas are usually accessible, the job of insulating them often isn't too tough. And the heat saved can be substantial.

If you are building a new home or adding to your present one, you should insulate between the floor joists over an unheated area before you nail down the subfloor. This way you can staple insulation with an attached vapor barrier, which must face upward, to the joists with little effort (see drawing **42-B**). To insulate before the sub-

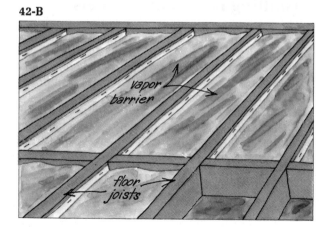

vapor barrier

floor joists

floor is nailed down, use the same techniques as those for insulating unfinished walls (page 40), sizing the insulation according to the information on page 12.

Mineral-wool batts or blankets with attached vapor barriers are also usually the best choices for insulating between floor joists where you can't work from above. Precut 4-foot batts are easiest to handle where space is limited. Where space to maneuver isn't a problem, use blankets, cutting them to the exact lengths needed for each space.

If you can find them, "reverse-flange" batts are excellent for use under finished floors. With standard batts, attachment is difficult because the vapor barrier faces up, where you can't get to the flanges for stapling. And because the barrier faces up, the fibers drift down into your face as you work. Reverse-flange batts have a vapor barrier on one side and, on the other, a perforated kraft paper or foil facing with stapling flanges. The extra facing keeps fibers from drifting down on you, and the flanges make stapling under floors easy.

For a dirt crawl space, lay 4 to 6-mil opaque polyethylene sheeting on the ground, extending it several inches up walls and fastening it there with duct tape (see drawing **43-A**). Overlap

43-A

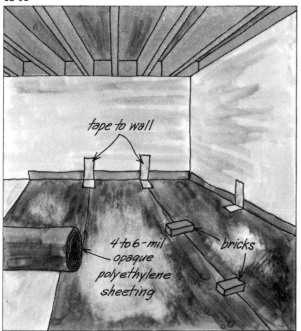

tape to wall

4 to 6-mil opaque polyethylene sheeting

bricks

adjoining pieces and anchor down the barrier with bricks or rocks. This will help keep ground moisture from being drawn up into the house. It also makes crawling under the floor a less dirty job and keeps insects and other pests out from under the floor.

Getting started

Here are some initial steps you should take before insulating under floors.

1) Check the floor joists to see whether you will need 16-inch-wide or 24-inch-wide insulation. Measure the length and width of the floor and compute the square footage you'll need, according to the information given on page 13. Figure the proper R-value, referring to that same page.

2) Read the discussion of tools and supplies on page 20, and gather the things you'll need. Plan to purchase anything you don't have when you buy the insulation. Don't neglect protection for your eyes, nose, and mouth.

3) Call around to a few lumberyards and building supply dealers to locate the best price for the insulation; find out if anyone has "reverse-flange" batts in stock. See if the dealer you choose will deliver the insulation. Then buy it and, one way or another, get it home.

4) Change into some old, loose-fitting clothes, gather your protective gear, and put all necessary tools in your tool pouch. Stack the insulation near the place you will start working. Get a piece of plywood or a large board that you can cut the batts or blankets on.

Cutting blankets and batts

Cutting a blanket is sometimes trickier than cutting a batt because of its length. But once you get the hang of it, cutting blankets and batts is easy. It's easiest to cut with the vapor barrier facing upward. To cut off a piece, just insert the blade of your utility knife shy of the far stapling flange. Cut toward yourself, right through the flange near you. As you cut with short strokes, spread the fibers with your other hand. Use your fingers to tear apart the far stapling flange. (See drawing **29-D** on page 29).

To cut a piece lengthwise, guide the knife along a scrap of lumber, such as a 1 by 4 (see drawing **30-A** on page 30). Allow about an inch extra in width so the insulation will fit snugly.

Fastening between floor joists

Because vapor barriers face up, you'll have to use methods other than stapling for fastening batts between joists, unless you can locate "reverse-flange" batts. Friction-fitting batts, made especially for use under floors, are easy to use, but they're also hard to find and don't offer a vapor barrier.

(Continued on next page)

Here are several methods for holding standard blankets and batts in place. For a couple of them, you'll need a hammer and nails; for all of them you'll need a pair of sturdy wire cutters.

Contractors use stiff pieces of wire cut to 15½-inch lengths. Each 4-foot batt is held in place by about three of these, bowed under the batts between joists. You can cut your own from 13-gauge wire or from old coat hangers. Make each one about ½ inch longer than the space between joists. (Look under "Wire" in the Yellow Pages to find 13-gauge stock.)

To install, just press batts or blankets into place between joists, vapor barrier facing up, making sure the adjoining ends abut. Fold up each batt or blanket at the ends of each joist space as shown in drawing **44-A**.

44-A

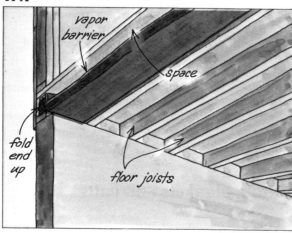

If there is room, leave an air space between the batt and the subfloor for additional insulating value. Bow a wire brace gently up against the fibers at one end. Put in an additional wire each 1½ feet (see drawing **44-B**).

44-B

If joists are not on standard spacings, use baling wire or chicken wire for support. Hammer flat-head nails into the bottom edges of joists, one every 2 feet, leaving ½ inch of each nail protruding. As shown in drawing **44-C**, you hold a batt

44-C

up in place between joists and lace 18-gauge baling wire back and forth from nail to nail. Or you can hold a batt in place between joists and staple chicken wire to the undersides of the joists (see drawing **44-D**). The baling wire method is usually easiest.

44-D

For stuffing cracks and places where the insulation does not fill, tear fibers from scrap pieces of insulation.

If you wish to protect the insulation from weather, nail panels of insulation board or standard interior-grade plywood to the undersides of joists. If you do this, you don't have to use the wire for bracing.

To clean up, throw your work clothes into the washing machine by themselves and take a cool shower. Unlike warm water, cool water will not cause the itchy insulation fibers to sink further into pores.

Insulating heated crawl spaces

If the crawl space under your house can be sealed up during winter, the cheapest and best way to insulate it is to drape insulation across the inner surfaces of the exterior walls and along the ground, as shown in drawing **45-A**. If you do this, you must also cover the ground with a vapor barrier of 4 to 6-mil opaque polyethylene.

Those who live in extremely cold climate areas, such as Alaska, Minnesota, or northern Maine, should not use the methods given here. Frost penetration can cause heaving of foundations. Consult your building codes or contact local HUD/FHA field offices for appropriate methods.

For insulating sealed crawl spaces, use mineral-wool blankets that have an attached vapor barrier. In addition to the usual tools, materials, and protective gear needed for insulating with blankets (see page 20), you'll need a hammer, some 10-penny nails, duct tape, and enough 1 by 2 and 2 by 4 lumber to stretch the combined lengths of walls to be insulated.

Here's how

For general techniques of cutting and working with blankets, see the information on insulating walls given on page 40.

Work along one of the walls running perpendicular to the joists first. Lay one polyethylene strip along the ground at the wall's base, using duct tape to fasten the strip to the walls at each end and along its length. (Refer to drawing **45-A**). Try not to walk on this vapor barrier very much as you work.

Next, precut a bunch of 1 by 2s to fit between the floor joists. Cut blankets long enough to reach from the underside of the subfloor to the ground plus 2 feet, so they will trail across the ground.

Now start working in a corner. Push a blanket firmly up against the subfloor, vapor barrier facing you, and fasten it there with a 1 by 2 as shown in drawing **45-A**. Nail through the 1 by 2 and the insulation and into the joist.

Cover the entire wall, butting each successive blanket firmly against the one before it. Repeat the process for the other wall that runs parallel to it.

Do the same for the walls that run parallel to the joists, but cut the blankets just long enough to reach from the subfloor to the ground (don't add the extra 2 feet).

Place a 2 by 4 along the bend of the blankets that trail across the ground, to hold them in place. Then finish laying all vapor barriers, overlapping them at their joints and tucking them under the blankets. Also lay a blanket across the ground, vapor barrier facing up, along the bases of the walls that parallel the joists (see drawing **45-A**). Weight each of these down with a 2 by 4. Place a few rocks or bricks on the polyethylene to hold it down, and you're finished.

Wash your work clothes by themselves and take a cool shower to wash away any itchy aftereffects of the insulation.

45-A

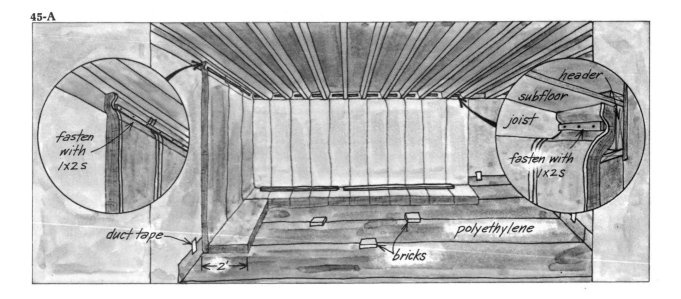

How to insulate heated basements

Because basement walls are usually made of concrete or concrete blocks, they present special insulating problems. You must provide a space for insulation and then, to cover up the insulation, wooden framing members that you can nail gypsum wallboard to. (Any wood-frame walls in a basement can be insulated by the methods used for regular walls, discussed on page 39.)

Of course, if you plan to finish the basement and use it as living or work space, insulating need not be a problem. Just combine projects and insulate as part of the finishing process.

The most difficult part of insulating a basement is the carpentry work involved in adding wooden framing members. Before insulating, you should weigh the possible savings against the difficulty of doing it yourself or the expense of having the job done.

If you can do the work yourself, it pays to insulate basement walls. But if you must have a carpenter do some of the preparatory work, you may save in the long term only if you live where winter fuel bills are a substantial problem.

Don't use the methods shown here if you live in Alaska, Minnesota, northern Maine, or another area where frosts are extreme. Penetration of frost can cause heaving of the foundation. Contact your building department or local field offices of HUD/FHA for recommendations.

What is required to insulate concrete basement walls? You must either build a wood-frame wall just shy of each concrete wall or you must furr out the concrete walls with smaller wooden framing members.

If basement walls are not flat or plumb, or if you wish to use insulation that is thicker than 1½ inches, it is probably wisest to build wood-frame walls separate from the concrete walls. (Drawing **46-A** shows a typical wall.)

On the other hand, if your basement walls are reasonably flat and plumb, and if insulation no thicker than 1½ inches is sufficient, the furring-out method is better.

Early in the project you'll have to make these determinations and decide on the type of insulation you will use. Because of high R-values in relation to thicknesses, rigid foam insulation is often the first choice for basements. But beware: because of the flammability of rigid foam insulation and the vapor-barrier facings on regular blanket or batt insulations, these materials cannot be left exposed. You must cover them with gypsum wallboard or some other fire-retardant covering permitted by building codes.

Because building a wood-frame wall requires

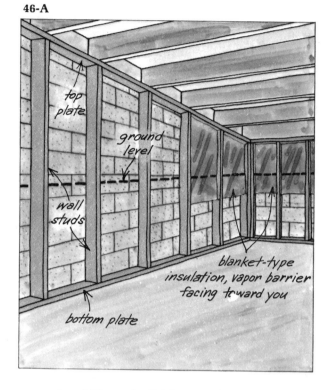

46-A

top plate

ground level

wall studs

blanket-type insulation, vapor barrier facing toward you

bottom plate

carpentry skills and a fairly complete assortment of carpentry tools, this subject is not covered in this book. For how-to-do-it information, see the Sunset book *Basic Carpentry* or consult a carpenter.

With a little guidance and a few tools, furring out a wall is not too difficult. Here's how to do it.

Measure and cut 2 by 2s to fit against the concrete wall, as shown in drawing **46-B**. Use a level to check the vertical lengths for plumb once they are held against the wall.

46-B

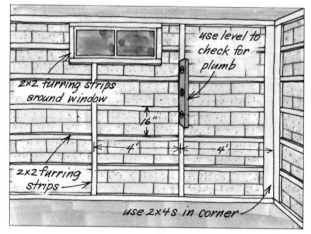

use level to check for plumb

2x2 furring strips around window

16"

4'   4'

2x2 furring strips

use 2x4s in corner

To attach the strips to the wall, use a caulking gun loaded with paneling adhesive. Follow the instructions on the adhesive tube. Where vertical members are not plumb or flat against the wall, shim them out with short pieces of shingle, as shown in drawing **47-A**.

47-A

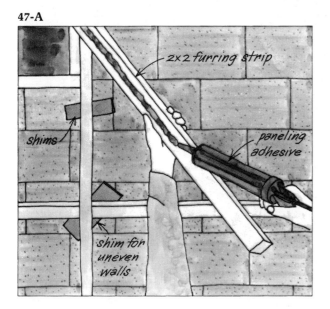

Lay the vertical members out so they are on 4-foot centers (so the long edges of 4-foot by 8-foot gypsum wallboard panels have wood backing). Run horizontal strips along the wall at top and bottom, and on 16-inch centers. Be sure they are level. Also apply furring strips around any doors or windows (refer to drawing **46-B**).

Unless you live in a cold northern climate, keep one fact in mind: you need only insulate the parts of the walls that project above ground level and down to about 2 feet below the ground.

Fit panels of insulation snugly between the furring strips, fastening them to the masonry with paneling adhesive. Use a utility knife or a handsaw to cut them to size.

If you can, insulate the perimeter of the joist spaces too. You can use rigid boards for this, but batts or blankets are easier to install there.

Face the vapor barrier toward you. As shown in drawing **47-B**, staple pieces in place between joists, overlapping the vapor barrier onto the furring strips and stapling to the sides of joists. Run a long piece along the box joist as well, overlapping it the same way. Staple one flange to the underside of the subflooring and one flange along the horizontal furring strip. Cut and handle this material as described on page 28.

If the rigid board has no vapor barrier, staple 2-mil polyethylene across the entire surface of the insulated walls, overlapping joints and repairing any tears with tape.

47-B

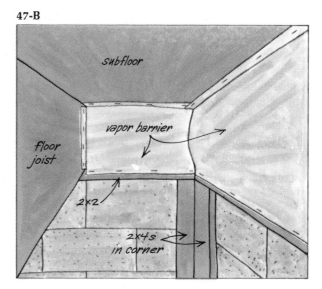

Last, but not least, cover all the insulation with ½-inch gypsum wallboard or another approved fire-retardant wall surface material. When you are nailing it in place, wedge it up off the floor about ½ inch so it can't be damaged by moisture (see drawing **47-C**). For information on taping joints and nailheads in gypsum wallboard, see "Closing the access holes," page 37.

47-C

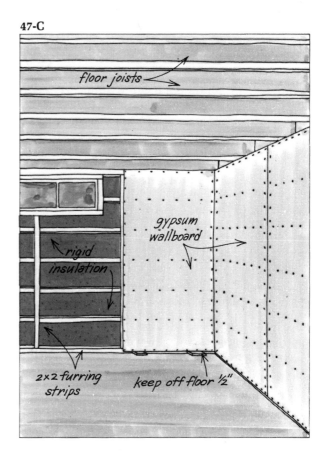

Insulating ducts and pipes

Where heating ducts run through unheated parts of a house, such as an attic, garage, or basement, they should be insulated. Ducts are rectangular or cylindrical sheet-metal passageways that carry heat from the furnace to your rooms. Unless they are insulated, they can waste more heat than they deliver. That means you pay for much more heat than you get.

Some water pipes need insulation too, especially cold-water pipes exposed to freezing temperatures in winter and hot-water pipes between the water heater and fixtures. Insulation can keep pipes from freezing and minimize heat loss through hot-water pipes.

If you insulate both ducts and pipes, you may reduce your annual heating costs by as much as 10 percent.

One note about water heaters: you can buy special insulation kits for either gas or electric water heaters. An insulated water heater retains from 5 to 12 percent of its energy that is otherwise wasted. Follow package instructions for installation. Be sure to keep the vents of a gas water heater free of all combustible materials.

How to insulate ducts

You can buy 1-inch or 2-inch-thick blanket insulation made for wrapping heating ductwork. Most building supply stores carry rolls of it; if you can't find it there, try a heating supply outlet or a heating contractor.

Blankets for heating ducts have no vapor barrier; for insulating air-conditioning ducts, buy blankets that have a vapor barrier and face the barrier outward.

For information on the tools and gear you'll need for working with insulation blankets, see page 20. In addition to the materials discussed there, you'll also need a roll of duct tape.

Before you insulate ducts, seal up any cracks, holes, or gaps with duct tape.

Insulating a cylindrical duct. One-inch-thick blankets for round ducts are usually about a foot wide. To insulate a round duct, start at either end. Tape the end of the blanket to the end of the duct, where it disappears into a wall, floor, ceiling, or furnace. Wrap the duct in a spiral fashion, overlapping each successive layer by half (see drawing **48-A**). If the whole roll won't pass between the duct and joists or framing, cut short sections that will. Tape ends of adjoining sections together, and tape the end piece to

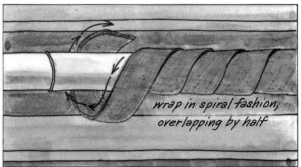

48-A

wrap in spiral fashion, overlapping by half

itself. Be sure that the entire duct is covered with insulation.

Insulating a rectangular duct. To insulate a rectangular duct, use 2-inch-thick blankets. Cut short sections that will each make one complete wrap around the duct. If the duct is recessed between joists, cut the pieces long enough to wrap as completely as possible, and staple the ends to the joists (see drawing **48-B**).

48-B

recessed duct

cut, fold, and tape at end of duct

staple to joist

Tape adjoining blankets together and, at the end of the duct, cut and fold the blanket as shown in the drawing.

How to insulate pipes

For insulating pipes, use special sleeves or self-adhesive pipe-wrapping insulation made for the purpose. A sleeve consists of an inner insulative material that wraps around the pipe and an outer covering that holds it in place (drawing **49-A**). The pipe-wrapping insulation is packaged in a coil that you wrap around the pipes like tape (drawing **49-A**). Some have a foil backing.

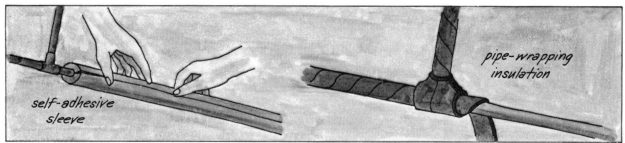

self-adhesive sleeve

pipe-wrapping insulation

Before putting on the insulation, clean dirt and rust from the pipe. In addition to insulating the pipes, wrap the fittings completely with insulation tape. For added reflective insulating value, you can wrap the insulating sleeves with aluminum foil.

Other tips for saving energy

In addition to the energy-saving measures this book focuses on—insulation, weatherstripping, caulking, and storm windows and doors—you can take a host of steps to cut your energy bills and save valuable resources. Following is a helpful checklist.

Lighting. Turn off lights when not in use, and don't use bulbs of higher wattage than necessary. Where possible, replace incandescent light fixtures with fluorescent fixtures—they consume less energy.

Winter heat. Keep doors and windows closed. Set the thermostat at 68° or lower in the daytime and 60° or lower at night. The present world energy situation (limited availability and high cost) calls for keeping temperatures as low as possible. By turning the heat down 10° at night, you can reduce your heating energy by about 10 percent. By turning it off at night, you can avoid the expense and discomfort of having the furnace cycling on and off. Consider installing a thermostat that turns down the heat automatically when you retire at night (or during working hours) and turns it back up just before you get up in the morning (or before you return home).

Tune up your furnace every few years, and change the furnace filters at least twice a year. Dirty filters make a furnace work harder than necessary. Keep heating vents and registers unobstructed. Insulate ducts where they pass through unheated areas.

Open drapes on sunny days and close them at other times. And be sure the fireplace damper is closed when nothing is burning in the fireplace, so heat won't escape up the chimney.

Summer cool. Keep drapes closed during sunny periods. Light-colored drapes are best—they reflect the heat. Shade windows on the south and west sides of the house with trees, overhangs, awnings, shades, or curtains.

If you have air conditioning, set the thermostat at the highest setting you can tolerate—78° to 82°. If you leave for part of a day, turn it even higher.

Appliance conservation. Adjust the water heater to no higher than 140°F (120° is better). Keep hot water leakage to a minimum; repair any leaky hot water faucets. Don't use more hot water than necessary when washing dishes or clothes.

Use as little heat as possible when cooking. Turn off cooking elements when not in use. Don't boil unnecessarily large amounts of water or use pots and pans that are larger than needed. And don't open the oven door unnecessarily— the oven loses 20 percent of its heat each time the door is opened. Never heat the kitchen with the range or oven.

Operate dishwashers and clothes washers with full loads. When you use a dishwasher, set it on "fast dry" and open the machine before the cycle is completed to let dishes air dry.

Defrost the refrigerator regularly, and don't open the door needlessly.

If your television set has an "instant-on" feature, it probably draws energy all the time— even when turned off. If you can't turn this feature off, bypass it by putting a switch in the cord or plugging the television set into a switch-controlled outlet.

If you have a microwave oven, use it for reheating or cooking in small quantities—it uses 30 percent to 70 percent less energy than a regular oven. But to cook a complete meal—if each item must be cooked individually—use the regular oven. For this type of cooking, a microwave oven will consume from 28 percent to 130 percent more energy.

Weatherstripping

Infiltration! Sounds devious, doesn't it? In most cases, it is. Infiltration—the free movement of air through openings in a house—can account for up to 35 percent of the total heating load in a properly insulated house.

Some infiltration is intentional and necessary: air should circulate through a house and vent outdoors. But proper ventilation is planned and controlled movement of air. In a typical unweather-stripped house, more than half the infiltration is uncontrolled and unnecessary. In winter, cold air pours in and warm air escapes through cracks and crevices around windows and doors and through small openings in walls, roof, and floors.

Weatherstripping is designed to block uncontrolled infiltration of air; it can reduce that 35 percent of total heating load to 15 percent. Some types of weatherstripping also repel wind-driven rain and moisture.

If you have a hard time picturing a substantial movement of heat through narrow cracks around windows and doors, consider this: an opening $\frac{1}{8}$ inch wide all the way around a door is equivalent to a hole about 8 inches square. Imagine an 8-inch-square hole wherever you have an exterior door and think how much wind it would catch on a cold day.

The cost of weatherstripping is relatively small, installation is easy, and resulting fuel savings are sizable—so consider weatherstripping a must for making your home energy-efficient.

Looking for existing weatherstripping

Your house may already have some weatherstripping. If it does, and if the weatherstripping provides adequate protection, you may have little need for this chapter. But some types of weatherstripping wear out quickly, and other kinds become maladjusted when doors or windows shift or swell in their frames (a common occurrence). So if your house has no weatherstripping, has some but not enough, or has full weatherstripping that is not effective, you should seriously consider upgrading your weatherstripping.

The first step is to determine whether or not your windows and doors have weatherstripping that works. To find out if it's there, look for it. If you don't know what weatherstripping looks like, refer to the information beginning on the facing page. Check around door jambs and heads; on top of thresholds or sills; and across door bottoms (see drawing **52-A**). And look for weatherstripping around windows (see page 70 to find out where weatherstripping belongs on windows).

If you find some, check its effectiveness. Where you can feel a draft entering the house, you know weatherstripping is needed. If, from inside, you

can see light shining through the cracks under or around doors or windows, you can be pretty sure that those areas need weatherstripping. Or you can test windows and doors with a candle flame—if the flame flickers as you move a candle slowly around the perimeter of a window or door, air is getting through.

In some cases, an adjustment is all that's necessary. Some types of weatherstripping are adjustable; others are not. Most adjustable types have elongated screw holes that allow you to move the weatherstripping slightly, once you've backed off the screws (see drawing **52-B**).

52-B

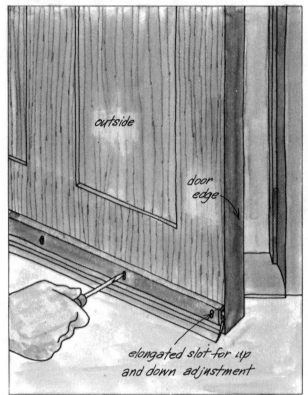

Though weatherstripping fastened with nails to a doorstop or jamb is not meant to be adjustable, you might be able to remove, adjust, and refasten it. Be careful not to kink or deform the weatherstripping.

Adjust weatherstripping with the door closed, fastening it as though you were installing new weatherstripping (see installation techniques, beginning on page 63).

If you find worn or deformed weatherstripping, peel or pry it off so you can replace it.

52-A

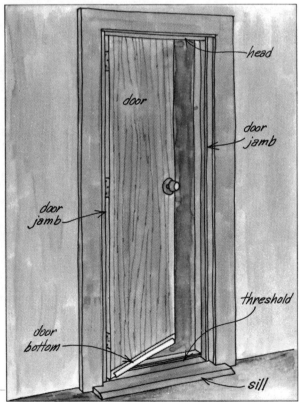

Types of weatherstripping

Weatherstripping is a catch-all word. Though it ordinarily refers to products designed to seal the cracks around doors and windows, the term can apply to practically anything that blocks air infiltration—including caulks and storm doors and windows (these are discussed later in this chapter).

For now, let's confine ourselves to what we normally call "weatherstripping products": materials designed to seal small cracks around doors and windows. The first part of this chapter focuses primarily on doors. For information on weatherstripping windows, see page 70.

For weatherstripping under doors, there are *thresholds* and *door bottoms*. Sealing the sides and tops of doors calls for *jamb weatherstripping*. Jamb weatherstripping is also used to weatherstrip double doors and some windows.

Those are the main types. In each category, a variety of products is available. We shall now take a closer look at them.

thresholds

A threshold fastens to the floor under a door. It helps to seal the air space under the door and hides the transition between one flooring material and another flooring material or the door sill.

To seal up the space under a door, a threshold works in conjunction with the door's bottom edge or a companion piece fastened to the door's bottom edge. Though most packaged thresholds are metal, you can buy one type—a "saddle"—made of hardwood.

When buying a threshold, be sure to get one long enough to reach between the door jambs at their widest separation. Several lengths are available, sized for standard doors.

Here are the main types of thresholds available:

Saddles

Also called "plain thresholds," most saddles are quite low and flat, beveled along both edges. Though a saddle does little to stop drafts by itself, it helps make an effective seal when combined with a door shoe, door sweep, or automatic door bottom (discussed on page 56). You can get saddles in either wood or metal (see drawing **53-A**).

Hardwood saddles offer the warm, natural appearance that only real wood can achieve. They are good choices where metal thresholds would look out of place—particularly in older homes, or where a saddle threshold should match hardwood flooring. Though inexpensive, they wear out more quickly than their metal counterparts.

Metal saddles are the type most commonly stocked at building supply stores and lumberyards. Aluminum is the most popular; bronze and stainless steel can be specially ordered. Though the latter types will outlast aluminum, they are considerably more expensive.

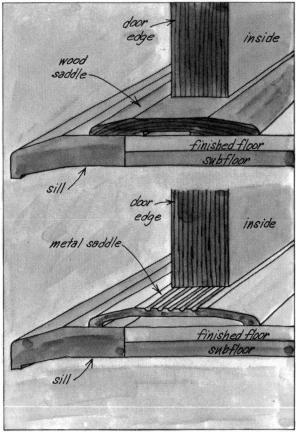

53-A

Vinyl-gasket thresholds

As shown in drawing **54-A**, a vinyl-gasket threshold works in combination with the underside of a door to seal up the space. A vinyl ridge, fitted in a groove running across the threshold's top, presses against the bottom of the closed door.

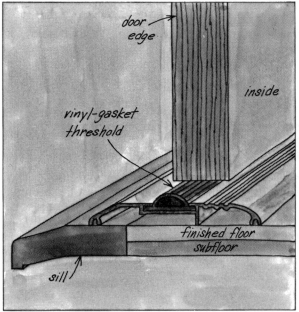

54-A

When in good condition, vinyl gasket is effective in keeping out dust, drafts, and water. Unfortunately, foot traffic eventually wears out the vinyl, lessening the seal's effectiveness. Because of this, most brands offer replaceable vinyl inserts. Before you buy, be sure your dealer stocks—or can order—replacements.

Interlocking thresholds

An interlocking threshold consists of two parts: the threshold itself, which fastens to the floor and to the door sill, and a "hook strip" that fastens to the door's bottom. As shown in drawing **54-B**, the two pieces interlock.

Though an interlocking threshold works very effectively when in good condition, it is more difficult to install than most other types and is easily damaged. If ice or gravel lodge in the channels, or if the hook strip gets bent, the door is difficult to close and the seal is poor. Interlocking thresholds are not recommended for use in cold climates where ice may form in the channels.

54-B

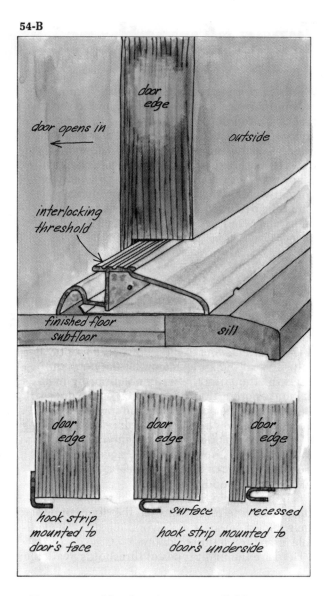

Two types of hook strips are available: one fastens to the bottom of a door's face, the other fastens to its underside. The latter type can either be nailed flat against the underside or recessed in a groove.

The face-mounted type is relatively easy to install and adjust. But because the hook strip screws directly onto the face of the door, it is more visible.

A hook strip fastened on the door's underside is not in view (see drawing **54-B**). But periodic adjustment is difficult. And cutting the groove for recessing it requires special tools and precise cutting. Unless you're a woodworker, this job is best left to a weatherstripping contractor.

To find a contractor, look in the Yellow Pages under "Weatherstripping Contractors." Most charge $40 to $50 per door to install a complete interlocking system (including door jambs as well as threshold).

Special thresholds

For some situations, ordinary thresholds won't quite do the job. Special thresholds are made for some of these purposes. Here is a discussion of three types.

Insulated thresholds. In extremely cold weather, household moisture condenses on cold metal surfaces, including the interior side of a metal threshold. A surprising amount of water may collect, eventually running off the threshold and damaging the floor. To prevent this, you can install a metal threshold with a plastic or vinyl separation to insulate the exterior side from the interior side (see drawing **55-A**). Insulated

55-A

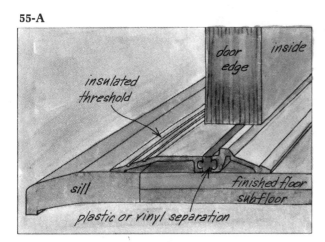

thresholds are practical where winter temperatures often drop well below freezing.

Waterproof thresholds. Some conventional metal thresholds have waterproof features built in. When combined with a rain drip (see page 57) attached to the door, they keep water from getting under the door. Any water that gets past the rain drip is caught in a water-return trough

55-B

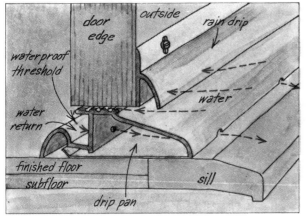

(drawing **55-B**) that drains it back outside, over the sill.

Half thresholds. Half thresholds are installed where two floors of differing levels meet. You can either buy a metal half saddle to serve this purpose (see drawing **55-C**) or make one by cutting off one of the beveled sides of a hardwood saddle (drawing **55-D**).

55-C

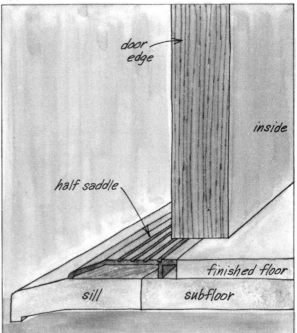

55-D

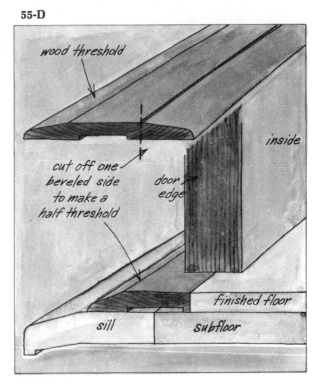

A "door bottom" is weatherstripping that attaches to the lower edge of a door. Several types are available. The most common kinds keep out drafts by pressing across the top of a threshold when the door is closed. These are called *sweeps, shoes,* and *automatic door bottoms.* Some of them seal out water as well as air. One slightly different variety of door bottom—a *rain drip*—sheds rain but doesn't seal the space under the door. Garage doors on heated garages should be weatherstripped too. Special sweeps and shoes are made for garage doors.

Following is a more detailed look at the various types of door bottoms. When buying any of these, be sure it's long enough to fit from one door jamb to the other (measure the side of the door on which you'll fasten it).

Door sweeps. A door sweep normally consists of a metal strip fastened to a felt, vinyl, or neoprene "sweep." It is simply screwed onto the bottom of a door (see drawing **56-A**).

56-A

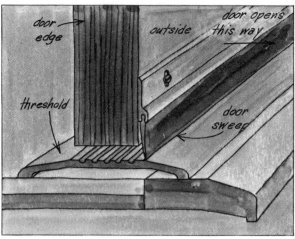

Sweeps are most commonly used on the exterior side of outward-swinging doors or on the interior side of inward-swinging doors that swing well above the interior floor. Because it is placed on the face of a door, a sweep is quite noticeable.

But sweeps are low-priced and easy to install—truly simple solutions to sealing out drafts. If you want sweeps that resist water, buy the vinyl or neoprene ones. In buying a sweep, pick one with slotted screw holes so it can be adjusted periodically. Sweeps come in several lengths, sized for standard door widths. Be sure you get one that's long enough.

Door shoes. A door shoe attaches to the underside of a door (see drawing **56-B**). Most types consist of an aluminum retainer holding a rounded vinyl ridge, much like the vinyl inserts in vinyl-gasket thresholds. You screw the aluminum retainer to the door's bottom edge; the vinyl ridge presses down against the threshold when the door is closed.

56-B

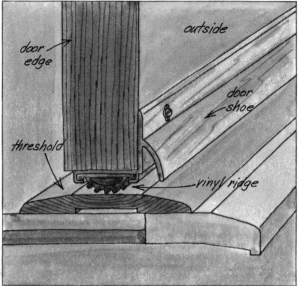

Though door shoes work over most saddle-type thresholds, you can buy thresholds made specially for door shoes.

Because the vinyl insert is under the door where it won't get stepped on, and because most door shoes can be adjusted, a threshold-and-door-shoe combination is usually more efficient than a vinyl-gasket threshold.

Door shoes, like sweeps, are available in lengths sized to fit standard doors. If you buy one, be sure it's long enough.

Automatic door bottoms. An automatic door bottom, sometimes called an "automatic sweep," is quite sophisticated. It consists of a felt or neoprene sweep spring-loaded inside a metal frame. The frame attaches to the door along the bottom edge. When the door is closed, the door jamb forces the sweep down against the threshold or floor. As the door is opened, a spring retracts the sweep into the frame. (See drawing **57-A**).

Automatic door bottoms work best directly over a door sill or floor, or in conjunction with low, flat thresholds.

57-A

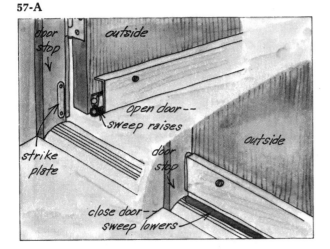

57-B

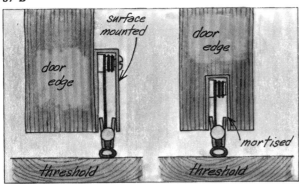

57-C

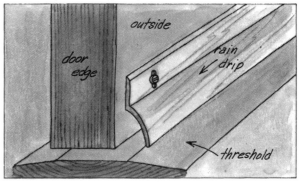

several lengths. You simply cut one to size and screw it along the lower edge of the door.

Garage door weatherstripping. Especially if the garage is heated, swing-up and roll-up overhead garage doors are usually weatherstripped across the bottom and along the sides. You can buy sweeps and shoes made specially for garage doors (check building-supply outlets or garage door companies). These sweeps and shoes are usually made of heavy-duty neoprene, rubber, or vinyl; sometimes these materials are clinched in an aluminum strip.

Most flexible types are nailed across the bottom of a garage door with ordinary roofing nails. The rigid aluminum type is screwed on in the same manner as ordinary door sweeps. Three types are shown in drawing **57-D**. Some sweeps

57-D

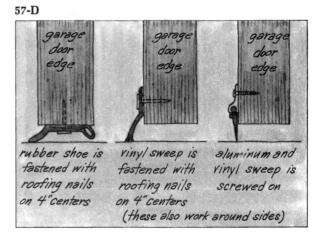

may also be used to weatherstrip the sides of a garage door.

If you have a leaky old garden hose, you might be able to substitute it for commercial garage-door-bottom weatherstripping. Just cut the hose to length and nail it along the door's bottom edge. (First be sure it isn't so thick that it won't let the door close all the way.)

Before you buy commercial garage door weatherstripping, be sure to measure the amount you need.

Two kinds are available (drawing **57-B**). One type is mortised underneath the door, hidden from sight. Installation of this type requires some expertise in woodworking. The other kind is surface-mounted, like a regular door sweep.

Though the surface-mounted type may be visually obtrusive, it is easy to install and does the job of weatherstripping as well as the recessed type. Because it is so visible, a surface-mounted automatic door bottom is usually mounted on the exterior side of an inward-swinging door. It is a good answer to the problem of weatherstripping the bottom edge of a door that swings over a floor or carpet slightly higher than the threshold (or sill).

Like other door bottoms, automatic ones are available in several lengths. When buying one, get the length that is just right—or *slightly* too long—to fit between the jambs. Though most kinds can be trimmed down slightly, cutting off too much of the length can damage the spring mechanism.

Rain drips. A rain drip sheds rain from the bottom of a door that swings over an exterior threshold. Some sweeps and shoes have a built-in rain drip, but you can also buy rain drips separately (see drawing **57-C**). They are available in

The many products that seal around the tops and sides of doors and around windows make up the category of jamb weatherstripping. Jamb weatherstripping is of three basic types: 1) spring metal and cushion metal; 2) gasket; and 3) interlocking. The first two types are easy to install with ordinary household tools. Following is a more detailed look at specific types.

Spring metal and cushion metal

Creating a seal by compressing between a jamb and a door when the door is closed (see drawing **58-A**), spring-metal and cushion-metal weatherstripping may be made of bronze, aluminum, or stainless steel.

58-A

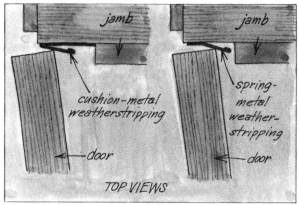

Spring metal is generally sold in rolls and attached to the jamb with small brads or nails. Cushion metal, attached in a similar fashion, comes as rigid V-shaped strips.

Though they provide an efficient seal against air leaks and are concealed when the door is shut, spring metal and cushion metal sometimes make a door slightly noisier and more difficult to open and close. And under some circumstances, wind may cause them to vibrate or hum.

Spring-metal and cushion-metal strips are likely to wear out or get torn because they depend on a rubbing action, but they usually last longer than similarly priced pliable-gasket weatherstripping.

Most spring-metal and cushion-metal weatherstripping can't be easily moved once in place, but you can adjust the seal by increasing the angle of the bend along the metal ridge.

In addition to their use for weatherstripping

doors, both types can seal double-hung windows and astrigals between double doors.

Gasket weatherstripping

Most jamb weatherstripping is of the resilient-gasket type. Typical varieties use resilient materials such as felt, vinyl, neoprene, or polyurethane foam to form a gasket-type seal between a door and its jamb.

You can buy either pliable gaskets, sold in rolls, or rigid gaskets, packaged in rigid strips sized to fit standard doors.

Pliable-gasket weatherstripping is usually less expensive than rigid-gasket, though most pliable kinds are not adjustable and are less durable. Various types are shown in drawing **58-B**. Polyurethane-foam tape, the least expensive jamb weatherstripping, is not recommended for exterior doors; it wears out quickly and can be difficult to remove entirely for replacement. Like other gasket weatherstripping, though, it works well to shut out light, dust, and drafts. Foam tape serves a variety of purposes, from weatherstripping windows and the cracks around air-conditioning units to sealing up automobiles.

Several kinds of felt are made for weatherstripping purposes, coming in rolls of varying widths and thicknesses. Wool felt, the most expensive, outlasts the lower-priced hair and cotton felts. Some types come with an adhesive backing that

58-B

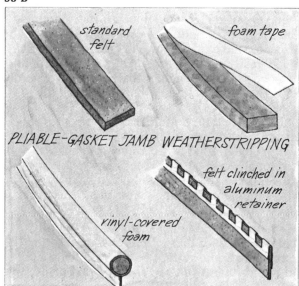

PLIABLE-GASKET JAMB WEATHERSTRIPPING

you press in place; others are nailed with rust-resistant carpet tacks or small nails.

The most durable pliable-gasket weatherstripping consists of vinyl, felt, or wool, clinched in a flexible aluminum retainer. Though typically more expensive than the others, this type will last considerably longer. During installation, you must be careful not to kink the material (see installation of this type on page 68).

Rigid-gasket weatherstripping, though more expensive than pliable-gasket, offers a higher-quality, more durable seal for exterior openings. Most types consist of a gasket material (vinyl, felt, or neoprene) attached to a metal or wood strip (drawing **59-A**). Some are extruded entirely

59-A

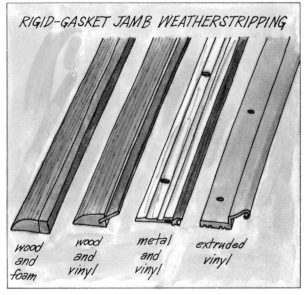

RIGID-GASKET JAMB WEATHERSTRIPPING

wood and foam   wood and vinyl   metal and vinyl   extruded vinyl

from vinyl. The strip is screwed or nailed to the door jamb. Some have elongated screw holes; these are good choices because you can adjust them periodically.

Before you buy, check the screws. Poor-quality screw/nails are difficult to loosen for slight adjustments of the strips. The heads of these screw/nails are often poorly made, with shallow or irregular slots.

Though rigid-gasket weatherstripping is quite visible, it offers the appearance of quality. You can paint the retainer strips, but be careful to keep paint off of the gaskets.

Interlocking jamb strips

Interlocking jamb strips consist of two parts: one fastens to the door's edge, the other to the jamb. When the door is closed, they interlock as shown

in drawing **59-B**, effectively sealing the door against wind and rain.

You can buy either the concealed or the surface-mounted type. Of course, the concealed

59-B

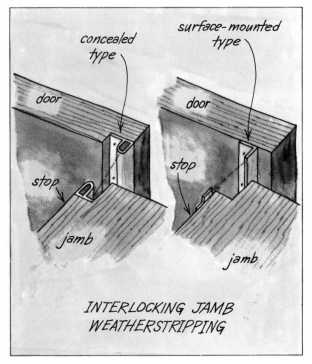

concealed type   surface-mounted type

door   door   stop   stop   jamb   jamb

INTERLOCKING JAMB WEATHERSTRIPPING

type is much less obtrusive than the surface-mounted type; it is recessed into grooves as shown in drawing **59-B**. Unless you are handy with tools, however, this type is best installed by a carpenter or professional weatherstripper.

The surface-mounted type doesn't require special cutting but precise fitting and alignment are needed.

Interlocking jamb strips share the disadvantages of interlocking thresholds (see page 54). In addition, they are difficult to adjust.

Astragal weatherstripping

Double doors in homes usually swing in one direction only (they are "single acting"). The wood strip or rabbeted stop between these doors is called an "astragal." Weatherstripping that seals the gap between double doors is referred to as "astragal weatherstripping."

Some companies make weatherstripping specifically for astragals, but you generally must order these types specially. For most double doors, ordinary jamb weatherstripping will suffice. The best types to use and instructions for their installation are detailed on page 69.

Preparing doors for weatherstripping

To install some types of weatherstripping, you must take a door off of its hinges. This is a good time to correct any problems with the door. The most common problems are sticking and binding and improper alignment. A door that is out of kilter may not close once you weatherstrip it (depending upon the type of weatherstripping you use).

Of course, if a door is seriously out of alignment, warped, or nonfunctional, you must either replace it or get involved in a major repair job. For detailed information on fixing doors, refer to the Sunset book *Basic Home Repairs*. For minor adjustments, the following information will help.

Diagnosing the problem

The most common causes of door trouble are:
• Hinges that are loose or set improperly (hinge mortises too deep or not deep enough)
• Improper fitting or hanging when installed
• Warping, swelling, or shrinking of wood
• Settling or shifting of the house

Before you go to the trouble of repairing a door, determine if weatherstripping alone can correct the problem. If a door works freely but doesn't quite contact the door stop completely, you might be able to close up the gap by fastening rigid-gasket weatherstripping to the stop. Large cracks between a door and its jambs, caused by shrinking wood or poor alignment, can often be closed by installing spring-metal or cushion-metal jamb weatherstripping.

On the other hand, if a door sticks or binds, you will have to remove it and plane it, sand it, or adjust its hinges.

Locating the trouble

Before you can unstick a door, you must find out where it is sticking. To do this, slide a piece of wood 1/16 inch thick (or a nickel) between the door and jamb, and then move it across the top and down the sides. With a pencil, mark lightly on the door the places where your gauge hangs up. (The door should have a uniform 1/16-inch space on both sides and across the top.)

Drawing **60-A** will help you determine whether you can correct the problem at the hinges or whether the door must be planed or sanded.

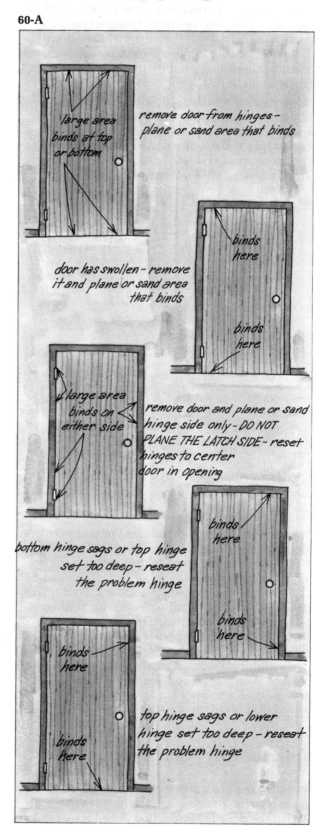

large area binds at top or bottom

remove door from hinges - plane or sand area that binds

door has swollen - remove it and plane or sand area that binds

binds here

binds here

large area binds on either side

remove door and plane or sand hinge side only - DO NOT PLANE THE LATCH SIDE - reset hinges to center door in opening

bottom hinge sags or top hinge set too deep - reseat the problem hinge

binds here

binds here

binds here

top hinge sags or lower hinge set too deep - reseat the problem hinge

binds here

Removing a door

To remove a door from its hinges, use a nail or nailset and a hammer to tap out the hinge pins, as shown in drawing **61-A**. Have someone hold

61-A

the door as you do this. Replace the pins in the leaves that remain on the jamb so they won't get lost.

Adjusting hinges

In many cases, a door may stick or bind simply because the hinge screws are loose, either in the door or the jamb. To remedy this, remove the door and the loose screws. For each screw hole, trim to size a wood peg (matchsticks will work), glue it, and pound it into the hole, as shown in drawing **61-B**. Then, cut it off flush, using a chisel. Replace screws.

If the hinges are fastened tightly but cause the door to hang crooked, you'll have to reset them. Depending upon the way the door binds, one hinge must be shimmed out and the other set

61-B

deeper into its mortise. Work on the hinge leaves that are on the door.

To cut the mortise deeper, shave wood from the bottom of the mortise, using a chisel, as shown in drawing **61-C**. Each hinge leaf should be just flush with the door's edge or slightly recessed. Don't set a leaf too deeply—instead, compensate by shimming out the other one.

61-C

Make shims from thin pieces of shingle, stiff (noncompressible) cardboard, or a similar material, as shown in drawing **61-D**. Or you can buy

61-D

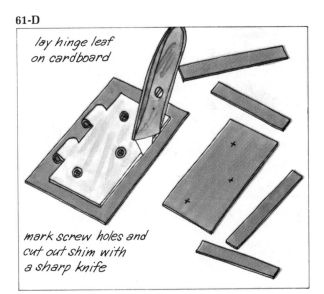

shims made from sheet brass. Brass shims can be purchased in most hardware stores in various sizes to fit standard hinges.

After installing shims, you may have to fill the screw holes or buy slightly longer screws to get them to bite securely into the wood.

Trimming a door

You may have to trim a door when weatherstripping, either to make room for the weatherstripping or to get the door to fit properly. Depending upon the amount of wood that must be removed, you can saw it, plane it, or sand it.

Don't be too hasty to plane. Coarse sandpaper followed by finer sandpaper gives you more control. For this type of work, a belt sander is easiest. If you don't have a power sander, wrap the sandpaper around a wood block to keep the sanding as even as possible.

Sometimes you'll have to plane a door to make it fit. Concentrate first on the hinge side. Avoid planing the latch side if possible, since it is beveled for a tight fit. Planing that side can also interfere with the fitting of the lockset. If you can't plane the hinge side to unstick a door, try planing the top or bottom.

About the only time you will need to cut a door with a saw is to make room for a threshold or door bottom.

How to cut off a door. A power circular saw with a combination blade works best for cutting off the bottom of a door. First mark the door bottom for cutting, as shown on the facing page in drawing **63-A**. Be sure to allow for any necessary space required between door bottom and threshold (depending upon the weatherstripping manufacturer's recommendations).

Make a guide for the saw by clamping a straightedge across the door bottom, as shown in drawing **62-B**. To avoid splintering the wood when cutting across the grain, score the line with a utility knife, where it crosses the vertical members of the door, as shown in drawing **62-A**. Then cut off the door bottom, keeping the blade to the waste side of your mark.

If you don't have a power saw, you can use a handsaw to make the cut. Work carefully, using a

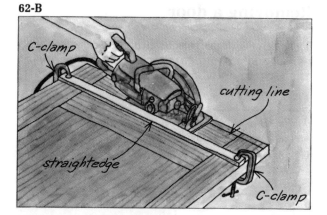

62-B

steady stroke. The saw guide shown in drawing **62-B** can also help you cut straight with a handsaw. When cutting by hand use a crosscut saw to cut across the door's vertical members, then switch to a ripsaw (if you have one) to cut in line with the grain.

Whether you use a power saw or handsaw, follow up with a few strokes of a wood rasp or plane to smooth out the cut.

How to plane a door. Here are some general rules for planing a door:
• Use a plane with a blade wider than the door's edge, in order to keep cuts flat.
• Keep your weight on the handle at the rear of the plane, not on the guide knob at the front.
• Make your strokes parallel with the edge you are planing, and keep the plane at the same angle as the direction of the stroke, as shown in drawing **62-C**.

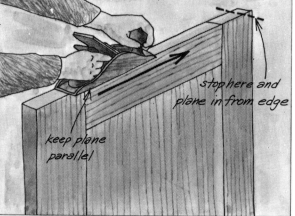

62-C

• When planing across the grain (across the ends of the side rails), work from the edges toward the center. Do not plane all the way over the edge. An end will often split under outward pressure from the plane.

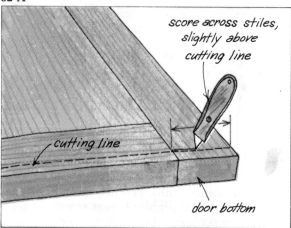

62-A

How to install weatherstripping

Installing weatherstripping is usually easy. In most cases, all you need are a few common household tools and the patience to align and fit the weatherstripping properly.

This section will tell you how to install the various types of commonly available thresholds, door bottoms, and jamb weatherstripping products—in that order.

installing thresholds

All thresholds are installed in the same basic manner: they are fit snugly between the jambs and fastened to the sill and floor with nails or screws.

Removing an old threshold

Old thresholds are easy to remove. To take out a metal one, simply unscrew it and lift it out. To remove a wood threshold, drive the nails all the way through, using a hammer and a nailset, then pry it out. Pull the nails from the sill.

When lifting or prying up a threshold, be careful not to mar or damage the finish floor or door jambs.

Trimming a door for a threshold

If you install a new threshold or replace an old one with a different kind, you may have to cut or plane the door's bottom edge to make room for the threshold.

To remove a small amount of wood (less than ¼ inch), use a plane or sandpaper. If you're removing more than that, cut off the door's bottom with a saw. To find out how to trim the door's bottom, see "Trimming a door," facing page. The following information shows how to mark the door for proper clearance.

Installing a wood threshold

The first step for installing a wood threshold is to butt the end of the threshold against the door and mark its level on the door at two or three locations, as shown in drawing **63-A**. From those marks, add the measurement for the suggested clearance and draw a line across the door's bottom. If you're installing a door sweep, door shoe,

63-A

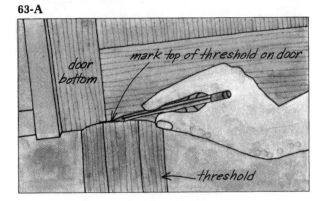

door bottom / mark top of threshold on door / threshold

or automatic door bottom, allow about ½ inch clearance between the threshold and the door's bottom (or follow manufacturer's recommendations). If you're not installing any of these, keep the space as narrow as possible, about ⅛ inch.

Remove the door and cut or plane the door's bottom to your line. Then cut (or notch) the threshold to fit between the jambs, as shown in drawing **63-B**. Use a coping saw or a portable

63-B

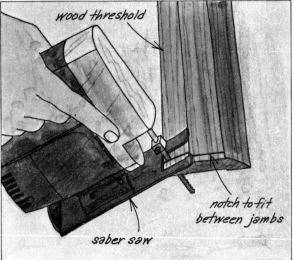

wood threshold / notch to fit between jambs / saber saw

saber saw for the cutting. Cut slightly to the outside of your lines—it's easy to trim off excess material, but adding on extra is another story. Use a wood rasp to remove excess wood and smooth your cuts.

Before you fasten the threshold to the sill, caulk underneath it and across its ends to seal out water (for information on caulking see page 73).

Fasten the threshold to the sill with 10-penny finish nails, centering the threshold under the door as shown in drawing **64-A**. Nail into the beveled sides of the threshold, not through the top. Because oak and other hardwoods split easily, it is best to predrill nail holes through the threshold and into the sill and floor, using a drill bit slightly smaller than the nail diameter. And don't nail too close to the wood's edges or ends: this also may cause splitting.

64-A

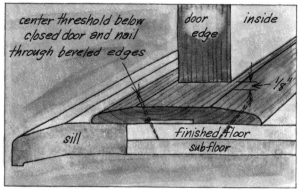

Set the nails and fill the holes with wood filler or pigmented putty. Then sand the threshold lightly and give it a coat of clear varnish or sealer. (Don't use paint: foot traffic would quickly wear it off.)

Installing a metal threshold

To install a metal threshold, first mark and cut off the door as you would for a wood threshold. To cut the threshold to fit between door jambs, you'll need a hacksaw with a fine blade. Use a metal file to trim and smooth it.

Some metal thresholds are equipped with a drip pan to funnel off collected water. This must be installed before the main part of the threshold. Caulk across the ends before you set it in place.

Center the threshold under the door, as shown in drawing **64-A**. Then fasten the threshold to the sill with screws. If the sill is oak, predrill screw holes to avoid splitting the sill and to make turning the screws easier.

Installing a vinyl-gasket threshold

Vinyl-gasket thresholds require about the same installation methods given for metal thresholds. Cut or plane off the door bottom so that it presses *lightly* against the vinyl gasket when the door is closed. Measure carefully before you cut. If you allow too much clearance, the seal will be ineffective. On the other hand, if you don't allow enough, the vinyl gasket will wear out quickly.

For a more effective seal, you can bevel the door bottom slightly, as shown in drawing **64-B**.

64-B

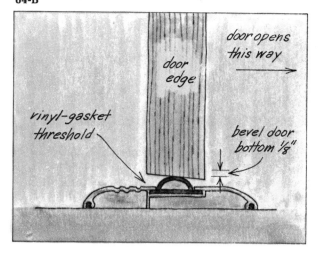

Installing an interlocking threshold

To install an interlocking threshold, follow the basic methods outlined for metal thresholds. Drawing **64-C** shows proper placement of an interlocking threshold under a door, where the threshold will align with a concealed or surface-mounted hook strip.

64-C

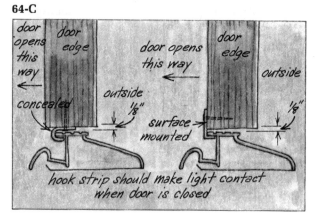

To install a concealed hook strip, you must take the door off its hinges. Cut or plane the door's bottom edge to allow ⅛ inch clearance between threshold and door.

Measure the width of the door and cut the hook strip to that length. Align the hook strip as shown in drawing **64-C.** For an airtight installation, run a light bead of caulking compound (discussed on page 73) between the hook strip and the door bottom before nailing the hook strip in place. Space nails about 2 inches apart.

To install a surface-mounted hook strip, first make sure that you've aligned the threshold with the door's bottom edge, as shown in drawing **64-C.** Allow about ⅛ inch clearance between the door and threshold. Install the hook strip with the door on its hinges—simply fasten the strip to the door bottom with screws.

how to install door bottoms

Door bottoms are meant to press against the threshold when the door is closed, sealing up the space under the door. (This is not true of rain drips, which just shed rain.) The main types of door bottoms are sweeps, shoes, and automatic door bottoms.

Two somewhat specialized types not discussed here are rain drips and garage-door bottoms. Page 57 shows how to install these.

Installing a door sweep

To install a door sweep, first measure the door's width between the jambs on the side to which the door swings. If necessary, cut the sweep to that length, using a hacksaw. In some cases, it's better to cut equal amounts off each end so the screw holes will be spaced evenly across the door's bottom.

With the door closed, position the sweep so the vinyl, neoprene, or felt touches the threshold lightly, as shown in drawing **65-A.** Hold it firmly in place and mark for screw holes. Then remove

it and use a drill or awl to start screw holes in the door. Finally, screw the sweep in place. Slotted screw holes allow you to adjust it if necessary.

Installing a door shoe

Two kinds of door shoes are shown in drawing **65-B.** The U-shaped one should slide easily onto the bottom of a standard 1⅜-inch-thick door.

65-B

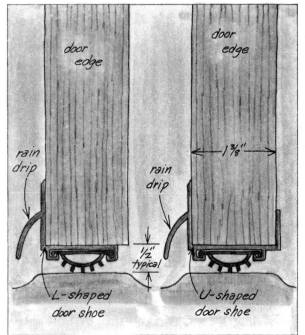

Both types are installed with the door on its hinges. If the shoe has a rain drip, you'll probably have to trim the ends to fit between the

65-A

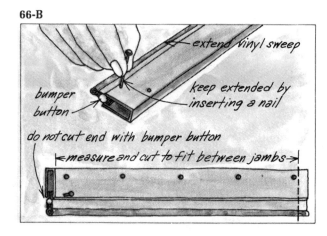

door stops before installation, as shown in drawing **66-A**.

To install a door shoe, first cut it to fit between the jambs on the side of the door you'll be fastening it to. For most door shoes, the best tool for this cutting is a hacksaw.

If necessary for clearance, take the door off its hinges and cut or plane the bottom as described on page 62. Fasten the shoe onto the door temporarily with a couple of screws. Then rehang the door and adjust the shoe so that the vinyl part of the shoe compresses lightly against the threshold when the door is closed. Mark for remaining screw holes, predrill them, and screw the shoe in place.

Installing an automatic door bottom

An automatic door bottom is installed in the same position as a plain door sweep. Measure the distance between door jambs on the proper side of the door. You will be cutting the automatic

door bottom to this length (subtract 1/16 inch if your type has a neoprene end on the push rod of the sweep and a small metal strike plate that fastens to the jamb at the point of contact).

Before cutting, depress the push rod to extend the sweep, as shown in drawing **66-B**. Keep it extended during cutting and installation by inserting a nail in the small hole provided for that purpose at one end.

Do the cutting with a hacksaw. Though most automatic door bottoms can be cut, don't cut off so much that you damage the inner mechanism; exchange it for a shorter one instead.

With the door closed, position the automatic door bottom so that, when extended, the sweep presses lightly against the threshold or floor. If the sweep exerts too much pressure when the door closes, the mechanism inside will soon wear out. Screw the automatic door bottom to the door. Then remove the nail to retract the sweep. If your type has a metal strike plate, fasten it to the jamb where the push rod will press against it.

how to install jamb weatherstripping

Jamb weatherstripping seals the small cracks around doors and windows. This section tells how to install it around doors. For information on using it for windows, see page 70.

The three categories of jamb weatherstripping are: 1) spring metal and cushion metal; 2) gasket; and 3) interlocking. Because interlocking weatherstripping is difficult for a homeowner to install, it is not discussed here. If you're interested in this type of weatherstripping, get in touch with a contractor.

To find a contractor, look in the Yellow Pages under "Weatherstripping Contractors." Most

charge $40 to $50 per door to install a complete interlocking system (including threshold).

Installing spring metal and cushion metal

Spring-metal and cushion-metal weatherstripping are installed on the door jamb next to the door stop, as shown in drawing **67-A**. With cushion-metal weatherstripping, the long leg of the V-strip butts against the edge of the stop.

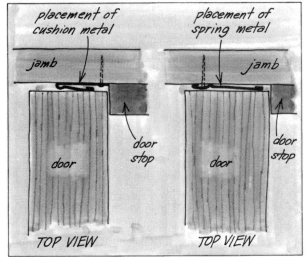

67-A

placement of
cushion metal

placement of
spring metal

jamb

jamb

door
stop

door
stop

door

door

TOP VIEW

TOP VIEW

After you've attached the side pieces, cut and install the top piece. Miter the corners, as shown in drawing **67-C**, so the strips will spring out

67-C

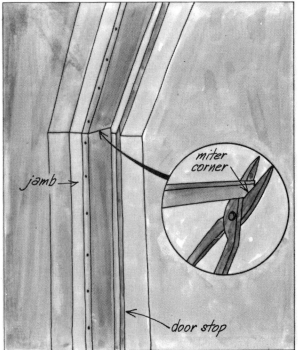

jamb

miter
corner

door stop

Spring-metal weatherstripping is positioned so that the edge doesn't quite contact the stop.

Before nailing along its length, tack each strip at its ends to align it. Spring-metal weatherstripping should be stretched flat against the jamb during installation so it won't kink or bow.

Starting with the hinge side of the doorway, cut the spring or cushion-metal strip with tin snips or a hacksaw and nail it to the jamb. Next, cut and install the strips on the latch side, above and below the strike plate. Some manufacturers provide a small strip with cushion-metal weatherstripping that fits behind the strike plate to complete the seal. Or you can make such a piece by cutting back the V-strip to fit behind the strike plate, as shown in drawing **67-B**. This strip should be installed first.

properly. When nailing, be careful not to bend or kink the strips (a tack hammer is handy for nailing).

"Set out" spring-metal strips by using an awl or icepick to score along the groove that runs their length, as shown in drawing **67-D**.

67-B

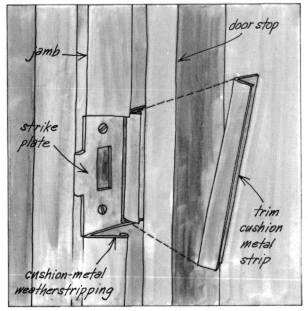

jamb

door stop

strike
plate

trim
cushion
metal
strip

cushion-metal
weatherstripping

67-D

score along groove to
"set out" spring metal

Installing gasket weatherstripping

Gasket weatherstripping is by far the easiest jamb weatherstripping to install. To attach adhesive-backed felt or foam, you simply press it in place where shown in drawing **68-A**. Tubular vinyl is

68-A

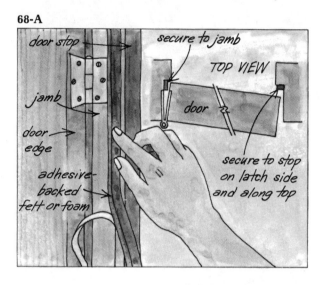

tacked to the stop so that it presses against the closed door (drawing **68-B**). Felt and vinyl types fastened to flexible aluminum nailing edges are positioned on the door stop the same way as tubular vinyl—but be careful not to kink the strip as you nail it.

Nail weatherstripping across the top of the doorway first, then down the side jambs. Pull

68-B

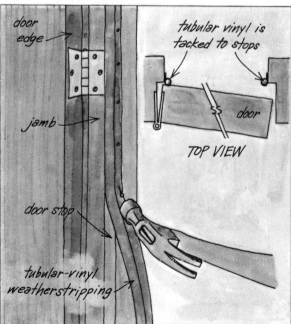

it taut as you nail. A tack hammer works well for nailing small, hard-to-hold nails. See drawing **68-C**.

68-C

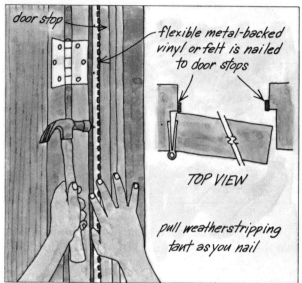

Rigid gaskets are equally easy to install. Cut the top strip to length with tin snips or a hacksaw and fasten it to the door stop, so that it presses lightly against the closed door (see drawing **68-D**). Then cut and install the side strips so

68-D

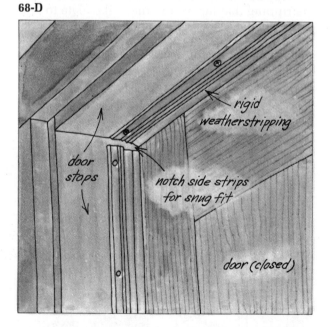

they fit tightly against both the top strip and the threshold. You may have to notch the top end of each strip to make it fit tightly, as shown in drawing **68-D**.

Installing astragal weatherstripping

Drawing **69-A** shows the plan views of a typical double door with a plain wood astragal, weatherstripped with felt and a spring-metal strip. The astragal is nailed to one door; the other door closes against it. For double weatherstripping

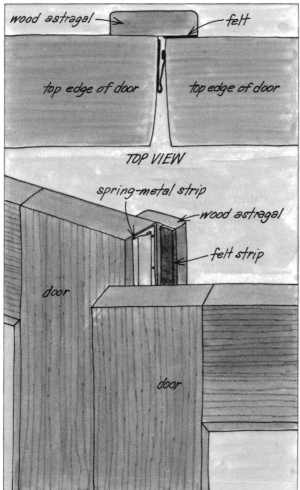

TOP VIEW

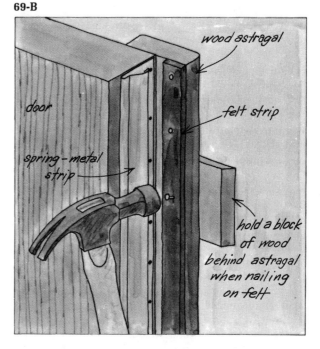

is in place, fasten the spring-metal strip to the door edge, following the instructions for fastening spring metal to jambs given on page 66.

Another option is to attach rigid-gasket weatherstripping with a vinyl ridge to the edge of each door so that the two will compress against each other when closed (drawing **69-C**).

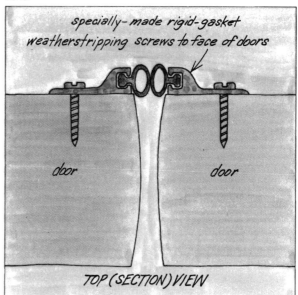

TOP (SECTION) VIEW

protection, the felt strip (about ⅛ inch thick) is nailed to the astragal where the door contacts it, and the spring-metal strip is fastened along the door's edge. Cushion-metal weatherstripping can be used, but it has a tendency to catch on people and things as they pass through the doorway.

Here's how to install the weatherstripping. First, nail the felt strip to the backside of the astragal, using a tack hammer and small nails or carpet tacks. Hold a block of wood behind the astragal while nailing to keep the astragal from coming loose (drawing **69-B**). Once the felt strip

If the door has an astragal, you can place the gasket strips on the side opposite the astragal. If there is no astragal, the strips can take the place of one.

Weatherstripping windows

Like doors, windows that open and close should be weatherstripped. Most windows manufactured in recent years are weatherstripped in the factory, but many older windows are not protected by weatherstripping. The majority of these can—and should—be sealed with ordinary jamb weatherstripping (discussed on page 58).

The types of jamb weatherstripping most commonly used for windows are spring metal and pliable gasket. Choice between these depends mainly on the type of window, the places where air leaks through it, and the importance of the window frame's appearance. Price and durability are also considerations.

For the sake of discussion, let's divide windows into a few basic types: double-hung sash, casement, sliding, and jalousie (louvered) windows.

Before you weatherstrip a window, be sure that it opens and closes easily and fits squarely in its frame. To find out how to repair windows, refer to the Sunset book *Basic Home Repairs*.

Double-hung sash windows

Spring-metal weatherstripping works best on double-hung sash windows. Though slightly more difficult to install than pliable-gasket types, spring metal is more durable. It is also hidden from view when the window is closed.

To install spring metal, raise the lower sash as far as it will go. Then measure from the bottom of the channel to a point about 2 inches above the bottom rail of the upper sash. Cut two strips to this length. Slide the strips into the lower sash channels, as shown in drawing **70-A**, so the nailing edges of the strips are flush with the channels' inside edges.

If the strips do not slide in easily, clean and sand the channels until they do. Nail the exposed portion of the strips, then close the lower sash and finish nailing the strips at the top.

Follow the same procedure for installing the upper sash channel strips. For most windows, you'll have to cut the spring-metal strips to fit around the upper sash pulleys, as shown in

70-B

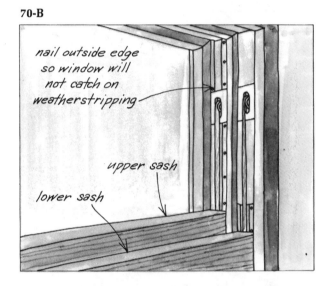

nail outside edge so window will not catch on weatherstripping

upper sash

lower sash

drawing **70-B**. As you fasten the strips, place an extra nail on the lower outside edge of the piece that is above the pulley, so the window won't catch on it when closing (see drawing **70-B**).

70-C

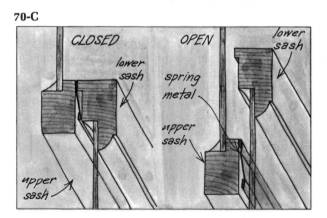

CLOSED   OPEN   lower sash

lower sash

spring metal

upper sash

upper sash

70-A

lower sash

insert 2" above the bottom of lower sash

spring-metal weatherstripping

lower sash channel

Next, cut strips to fit across the upper and lower sash rails and nail them where indicated in drawing **70-C**. Make sure the strips extend across the full width of the sash. After you've installed them all, use an awl or ice pick to "set out" the strips as described on page 67.

Pliable-gasket weatherstripping varieties used for double-hung windows include tubular vinyl and vinyl or felt clinched in a flexible aluminum retainer. Both kinds attach to the outside of the window and frame so they're not visible from inside.

To install, first cut pieces for the sides of the upper and lower sash. Nail them where shown in drawing **71-A**. The vinyl should compress lightly

71-A

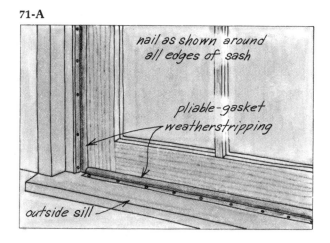

against the sash—if it fits too tightly, the window will be difficult to open and the vinyl will wear out quickly, breaking the seal.

Next, cut strips to fit across the rails of the upper and lower sashes (you'll need three pieces of equal length). Nail them on the sashes as shown in drawing **71-B**.

71-B

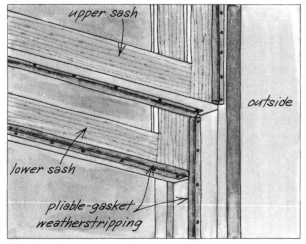

Casement windows

There are two main types of casement windows: metal and wood. Here's how to weatherstrip them.

Metal casement windows, as a rule, cannot be weatherstripped effectively with ordinary jamb weatherstripping. However, a spring-metal strip that simply clips onto the edges of the window is available. Illustrated in drawing **71-C**, these strips usually must be special-ordered from the manufacturer. If you are unable to obtain this material, you should consider storm windows that fasten to the inside of the casement frame as a permanent solution.

71-C

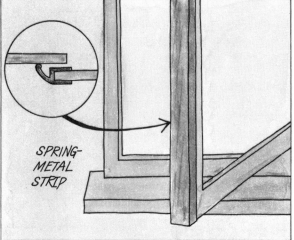

As a temporary solution, you can use adhesive-backed foam or felt strips for metal casement windows. The strips will soon wear out, though, and you may have difficulty in latching the windows. Attach the strips to the frame as shown in drawing **71-D**. Before you stick the strips on, clean the metal surface.

71-D

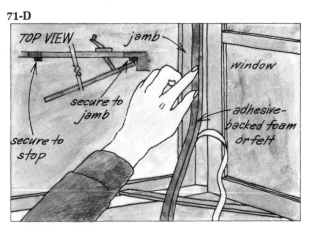

Wood casement and awning windows are weatherstripped just like door jambs and heads. Pliable-gasket and spring-metal or cushion-metal weatherstripping are most commonly used.

72-A

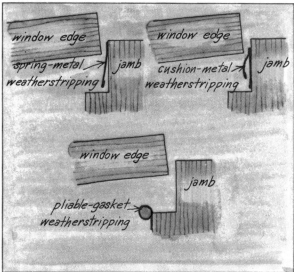

Drawing **72-A** shows where the weatherstripping goes. Refer to page 66, "How to install jamb weatherstripping," for further installation details.

Sliding windows

No weatherstripping is made specifically for home installation on sliding metal windows. The manufacturer's weatherstripping usually consists

72-B

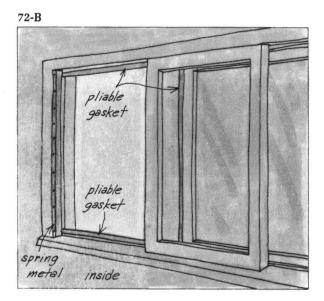

of a bead of felt, vinyl, or neoprene, gripped in a molded channel. It can be replaced only by a professional glazier or weatherstripping contractor.

Sliding wood-sash windows are easily weatherstripped with pliable-gasket weatherstripping (page 58). If both sashes on the window move, weatherstrip them as you would a double-hung window, following the directions given for pliable-gasket weatherstripping on page 68.

If only one sash moves, install pliable-gasket strips as shown in drawing **72-B**. Instead of attaching these strips on all four sides of the sash, however, you can fasten a more durable spring-metal or cushion-metal strip in the channel where the sash closes against the side of the frame.

Jalousie (louvered) windows

Though jalousie windows are often the smallest windows in your house, they can also be the draftiest. Air can leak around each individual pane, making these windows nearly impossible to weatherstrip.

One partial solution is to install a strip of clear vinyl across the bottom of each pane, as shown in drawing **72-C**. But this weatherstripping is hard

72-C

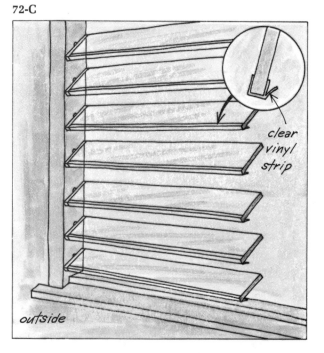

to find and must be ordered specially. The only other way to make jalousie windows weatherproof is to install a removable storm window on the frame during the cold season (see "Storm windows and doors," page 76).

Caulking your house

Even when doors and windows are weather-stripped, air can infiltrate through cracks and joints in the walls, roof, and floors of your house. Individually, these cracks and joints may seem negligible, but they combine to cause chilly drafts and raise heating costs. They also invite leaks, insects, and rot-causing moisture. Though you can't expect to seal up your house completely, caulking it properly can curb some of these problems.

Caulking is simple and relatively inexpensive, and it can cut up to 10 percent off your heating bills. You simply fill cracks, joints, holes, and crevices with caulking compound, applied with a caulking gun or putty knife. The process can pay for itself through savings in less than a year.

Where should you caulk?

A house is made of many different materials. With age, temperature extremes, and vibration, cracks usually develop where different materials meet—around window frames, for example. These are the places you should caulk. Drawing 73-A illustrates many of them.

Normally, caulking crevices is not necessary inside a house, but you may want to caulk in an unheated attic where wires, pipes, or vents go through the ceiling below. And you might want to caulk any cracks that provide access for insects. Spackling compound often works better than standard caulks for these types of jobs.

73-A

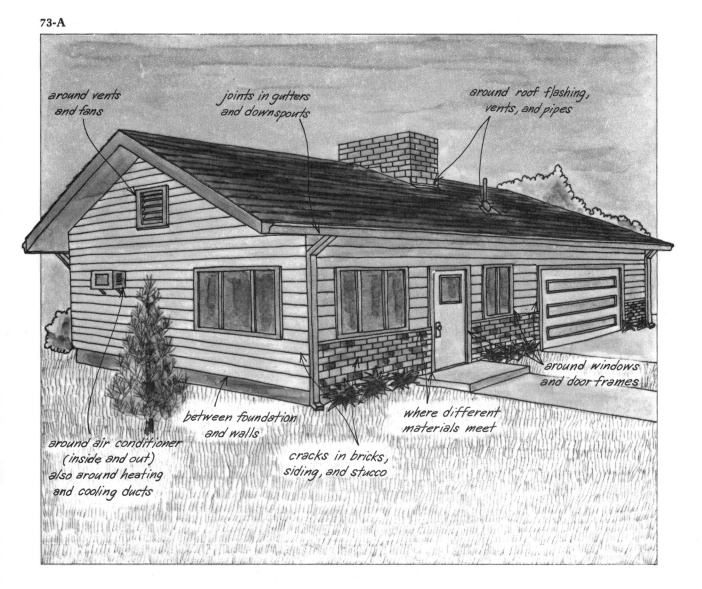

around vents and fans

joints in gutters and downspouts

around roof flashing, vents, and pipes

around windows and door frames

between foundation and walls

where different materials meet

around air conditioner (inside and out) also around heating and cooling ducts

cracks in bricks, siding, and stucco

You can choose from a broad spectrum of caulking compounds. They vary in both price and composition, and there is a direct relationship between the two. As a rule, you get what you pay for.

Elastomeric caulks

The best caulks are the elastomerics. Relatively new to the consumer market, these synthetic "supercaulks" effectively seal almost any type of crack or joint and adhere to most materials. They will outlast ordinary caulks by many years.

The generic types that fall into this category include polysulfides, polyurethanes, and silicones.

These products do have some drawbacks other than cost. Silicone rubber is hard to smooth out once applied, and it can't be painted over. Polysulfide can't be used on porous surfaces unless a special primer is applied first. As with other caulks, the surface on which you apply these should be clean, dry, and free from oil or old caulking material.

Latex and butyl-rubber caulks

You can also buy all-purpose caulks that offer average performance. These medium-priced products include latex, acrylic latex, and butyl rubber. Unlike the elastomerics, these are not pliable enough to use on large joints that expand or shrink (though some can withstand some movement and won't shrink away from the joint).

Latex and acrylic latex caulks vary in price and performance. The acrylic latex caulks outperform nonacrylic latexes. If the label doesn't say "acrylic," the product is probably the cheaper, nonacrylic variety.

Both latexes are easy to apply and clean up with water. Unlike other caulks, they can be applied to damp surfaces, though they are not recommended for surfaces that are continually exposed to water. They shouldn't be used on metal surfaces. Latex caulks can and should be painted to prolong their life.

Butyl-rubber caulk generally outperforms latexes. Though not considered a true elastomeric, butyl is more flexible and durable than acrylic latex. You can use butyl rubber on any type of surface or material. It has a tendency to shrink slightly while curing.

Oil-base caulks

Oil-base caulks are the lowest-priced, lowest-performance caulks on the market. Limit their use to stable interior cracks. For most purposes, you'll save money in the long run by selecting longer-lasting acrylic latex caulk.

Special-purpose caulks

Relatively new on the consumer market are polyurethane foam caulks in pressurized cans. Though expensive, these are particularly handy for filling large holes, cracks, and joints. Foam caulk comes out of the can as a sticky froth that adheres to most surfaces. As it cures, the substance expands to 2½ times its original volume and becomes semirigid. Be sure to follow label directions.

For sealing lap joints in sheet metal and corrugated metal roofing, and for joints in gutters or downspouts, use butyl gutter and lap seal.

Asphalt-base caulks are best for sealing chimney and vent flashings and for patching roofs. For patching masonry and concrete, use asphalt or rubber-base caulks.

Rope caulk, usually oil-based, is used as a temporary filler for very wide cracks or joints. To apply it, you simply unroll as many strands as you need and stuff it into the crack. Rope caulk is considered temporary because it won't adhere to the surfaces it contacts. Most rope caulks resist shrinkage very well and remain pliable. (In fact, you can generally remove and reuse them if you wish.)

For caulking between glass and window sash, use glazing compounds or linseed-oil putty. Before you apply them, remove any cracked or loose putty and prime the surface to be glazed with an oil-base paint.

Read the label!

Before you buy any particular type of caulk, read the label, including directions and precautions. Some caulks don't work in cracks or joints less than ¼ inch wide; others work well only in narrow cracks. Some types are flammable, irritating to skin, or give off dangerous vapors. Be sure you're aware of all these things before using the material.

Caulks come in two consistencies: knife-grade and gun-grade. Knife-grade caulks (including putty, spackling compound, and glazing compounds) are sold in cans and applied with a putty knife or spreading tool. Gun-grade caulks come in squeeze tubes for small jobs, in large containers for filling bulk-type guns, and—most commonly—in 11-ounce cartridges that fit into standard caulking guns (drawing **75-A**).

75-A

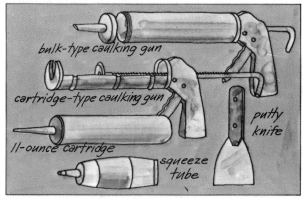

bulk-type caulking gun

cartridge-type caulking gun

11-ounce cartridge

putty knife

squeeze tube

For most caulking jobs, the 11-ounce cartridges and the caulking gun are the easiest to use. Using a bulk-type gun or a putty knife requires tool cleanup that sometimes calls for special solvents. The standard cartridge gun costs $3 to $5.

Preparation

Before you caulk, be sure the crack or joint is clean, dry, and free from oil, grease, or old sealant. With some caulks, narrow cracks or joints must be widened so the caulk will flow into them.

First scrape the crack or joint with an old screwdriver or chisel to remove old sealant and any other foreign matter. Then use a stiff brush to clean the crack further. If necessary, remove grease or grime with paint thinner or some other solvent. If the caulk requires a primer, prime all cracks and let the primer dry thoroughly.

Loading the gun

To load the caulking gun, turn the plunger so the teeth face away from the trigger, then pull the plunger back as far as it will go. Before you insert

the cartridge, remove any old sealant or matter from the gun. Drop in the cartridge and turn the plunger so the teeth face down toward the trigger. Then push the plunger until the teeth engage.

Snip off the end of the plastic nozzle at a 45° angle, as shown in drawing **75-B** (the size of the caulk bead will be determined by where you snip). Puncture the inner seal at the base of the nozzle with an icepick or long nail.

Applying the caulking compound

It may take a bit of practice to get the bead of caulk to flow evenly. Start by holding the gun at a 45° angle to the surface, as shown in drawing **75-B**. Then, as you move the gun across the sur-

75-B

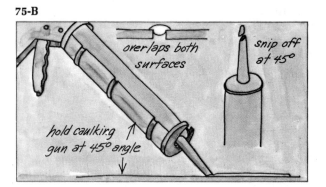

overlaps both surfaces

snip off at 45°

hold caulking gun at 45° angle

face, pump the trigger to keep caulk flowing uniformly. Make sure the caulking compound fills the crack completely and overlaps the adjoining surfaces.

After you've finished, disengage the plunger rod and pull it back slightly to keep the caulking compound from oozing out. If any caulk is left inside the cartridge, keep it from hardening by covering the nozzle with foil, masking tape, or plastic wrap.

Cleanup

If you've used any tools to apply caulking compound, clean them immediately after use, before the caulk has a chance to harden on them. It's a good idea to keep a solvent-soaked rag handy as you apply caulk. The appropriate solvent is usually mentioned on the label. For latex or acrylics, use water as a solvent.

Storm windows and doors

Windows and doors in a typical home can occupy as much as 40 percent of the exterior wall area. These windows and doors transfer expensively heated or cooled air much more rapidly than do walls. When it comes to saving energy, they clearly deserve close scrutiny. If you live in a climate where winters are blustery and cold, you should consider storm doors and windows.

How do they work?

Most heat moves through ordinary windows in two ways—by infiltration through cracks and open spaces around frames, and by conduction through the glass and framing materials. (For explanation of heat movement, see page 6.) Radiation, also a source of heat movement, heats up houses on hot days more than it cools them in winter.

Doors allow substantial transfer of heat too. Most heat movement occurs when you open and close them, but some results from the same type of infiltration and conduction that happens around windows.

Reduction of unwanted heat movement through doors and windows can be achieved in several ways. As we've already discussed, caulking and weatherstripping can greatly reduce infiltration. But neither of these methods reduces the large amount of heat transferred by conduction.

To reduce conduction of heat, you must insulate windows and doors. Though this may sound peculiar, it's not. The insulation for windows isn't fiberglass or rock wool—it's air or an inert (safe) gas. When confined, air and inert gases are excellent insulators.

A storm window or storm door creates a "dead air space" on one side of a window or door, reducing heat conduction by as much as 50 percent. See the chart below for R-value comparisons of typical windows and storm windows.

R-VALUES OF WINDOWS

Window Type	R-Value
Single glass	0.88
Double glass, 3/16″ air space	1.49
Double glass 1/4″ air space	1.64
Double glass 1/2″ air space	1.73
Double glass 3/4″ air space	1.82
Single glass with storm window	1.89
Double glass with storm window	2.65

Types of storm windows

Many products are used for creating dead air space around windows. Though all are termed "storm windows," some are merely raw building materials that you apply to windows. Some snap or stick onto the inside surfaces of window frames; others are fastened onto the outsides of frames.

Though the costs range from inexpensive to relatively expensive, keep in mind that all storm windows will repay their cost through energy savings—it's just a matter of time. The least expensive kind will more than pay for themselves in a year, but they are not durable enough to last any longer than that. Expensive types may take from 6 to 10 years to repay their cost, and they will last indefinitely.

Following is a brief rundown of some standard types of storm windows.

Polyethylene sheeting. Though at best, polyethylene sheeting will last only one season, installing it over windows is cheap, easy, and effective. Problems are that it obscures the view and doesn't allow the window to be opened for ventilation.

You can buy polyethylene either by the roll or in kit form. Measure your windows to figure the right size of rolls or kits to buy. Be sure to get 6-mil or thicker polyethylene that is transparent.

You can attach polyethylene sheeting to either the inside or the outside of a window. Inside, it lasts longer but looks somewhat unsightly. Outside, it is less conspicuous but exposed to damage by weather.

To install polyethylene inside, just cut it to size, using scissors, and tape it around the window with 2-inch-wide masking tape (see drawing **76-A**). Try to place it where the space between

76-A

tape around window with 2″ masking tape

polyethylene sheeting

inside

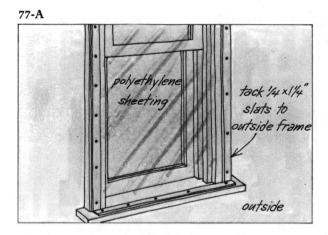

77-A

the window and the plastic will be ¾ inch or more. Stretch it tight as you tape.

If you choose to install polyethylene outside, secure it around the window's perimeter by tacking through wood laths or ¼-inch by 1¼-inch slats, as shown in drawing **77-A**. Again, try to keep a minimum space of ¾ inch between the plastic and the glass.

Single-pane kits. Many home improvement stores and some catalog department stores stock kit materials for making your own storm windows. The storm window material is usually rigid acrylic sheet plastic. Most types have mounting trim of vinyl, aluminum, or wood that frames the plastic.

You can get these storm windows for installation on either the inside or outside of windows. Some types that mount inside have adhesive-backed mounting trim that you simply press in place. Windows that mount outside usually come with pivoting clips to permit removal of the storm window for storage during the warm seasons (see drawing **77-B**).

77-B

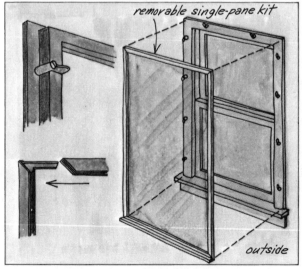

Though installation is quite easy (instructions should accompany the packages), the main problem with storm windows of this type is that you can't open them. Unless you have an easy escape route, this can be dangerous in the event of fire.

If you take these windows down after a cold season, be sure to mark the frames so you know which one belongs where.

Custom single-pane storm windows. For slightly more efficient, durable, and costly storm windows, you can have single-pane storm windows made to your measurements by a window supplier, then have them installed or install them yourself. The main problem with these, as with all single-pane storm windows, is that you can't open them. Again, this can be hazardous in the event of fire unless you have other escape routes.

This type of storm window usually goes on the outside of the primary window. Determine where the storm window will be placed (try to allow at least ¾ inch of dead air space). Measure the height and width as accurately as possible. Give these measurements to the supplier, and be sure to state that you have not figured in any allowances for discrepancies in fitting.

When you get the windows, check them carefully against your order. If they are not the proper sizes and do not fit, return them. For mounting to the window frames, use pivoting clips so you can take the windows down during the warm seasons.

Multi-track windows. You can buy "double-track" or "triple-track" storm windows for installation over operable windows. Both types are installed permanently to the outside of a regular window and can be opened easily for ventilation or escape during fire. Most kinds nail on. Triple-track windows have a screen that you can slide in front of the open section (see drawing **77-C**).

77-C

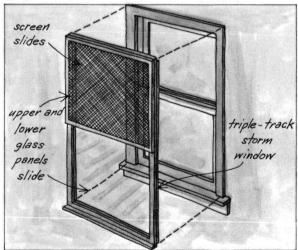

Several different finishes and styles are available. Be sure that the brand you order is well made and weatherstripped tightly around the channels and joints so air won't leak through.

Because double-track and triple-track windows are normally ordered through and installed by a contractor, you may have to wait several weeks for installation after measurements are taken.

Insulated replacement windows. Several companies offer specially designed insulated windows, sized to replace ordinary windows. These are especially appropriate choices if your windows are badly deteriorated or if you prefer low-maintenance windows and a long-term, energy-saving investment.

The glass part is either "double-glazed" or "triple-glazed." This means that two or three pieces of glass have been permanently fused around their edges, with a space—usually about 3/16 inch—between them (see drawing **78-A**). In

78-A

double-glazed window

space filled with air or inert gas

CUT-AWAY VIEW

some types, the space is filled with air; in others, the space contains an inert (safe) gas. The latter windows insulate more effectively than those filled with air.

Typical frame materials include wood, vinyl-clad wood, aluminum, and vinyl-coated aluminum. These windows are available in hundreds of sizes and many styles. All major varieties of windows are available: awning, casement, double-hung, picture windows, sliders, and so forth.

Some have a removable sash for easy cleaning from inside the house. Be sure to buy a brand that is weatherstripped at all joints. For dealers and installers of insulated replacement windows, look in the Yellow Pages under "Windows."

Storm doors

Many styles of storm doors are available that fasten directly to the door jamb, outside the primary door. They may have one glass panel, two glass panels, one glass panel and an interchangeable screen, or screens that slide behind the glass panels (see drawing **78-B**). They come in a variety of finishes and styles. All are effective if properly fitted and installed.

78-B

one glass panel

two glass panels

glass panel and interchangeable screen

self-storing screen

Unless you're an experienced do-it-yourselfer, it's usually best to have a contractor install a storm door, to insure that the job is done properly so that the door works well. If you tackle the job yourself, buy a packaged storm door from a building supply company and follow the instructions (they vary widely, depending upon the type and make of door).

Sunscreens

Bright sunlight streaming through closed, uncovered windows in summer can turn your home into a hothouse. Though storm windows and doors will block infiltration and conduction, radiation of heat is the culprit in summer.

To block sunlight's radiant heat, you can use any of several methods. For example, you can substitute a newly developed fiberglass sunscreen for your existing insect screening. This sunscreen blocks up to 70 percent of the radiant heat and cuts glare and reflections as well as keeping out insects. And although it allows plenty of light to enter the room, outsiders can't see in. It is available in some hardware stores and home-improvement centers (you may have to order it specially).

Metalized sheets of polyester film, applied to the inside of window panes, are quite popular for reflecting sunlight. Such films, made by several companies, are typically available in 36-inch and 48-inch widths. They are applied to the inside of windows like decals, usually with water and a squeegee. (You can either do this yourself or have a professional do it.)

The reflective sheen of these films bounces back up to 75 percent of the sun's heat and 82 percent of its glare. It also helps to check fading caused by the sun's ultraviolet rays. Some manufacturers claim that during winter this material can reduce heat loss through glass by 26 percent.

During daylight, you can see out through a comfortably tinted window, but outsiders can't see in. At night, however, this principle is reversed: the windows reflect on the inside.

To find dealers and applicators in your area, consult the Yellow Pages under "Glass Coating & Tinting." For cleaning, use nonabrasive cleaners with a soft cloth.

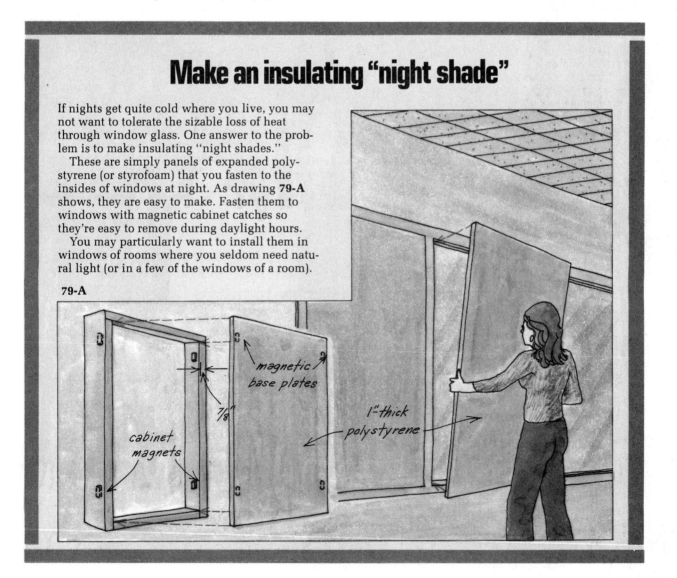

Make an insulating "night shade"

If nights get quite cold where you live, you may not want to tolerate the sizable loss of heat through window glass. One answer to the problem is to make insulating "night shades."

These are simply panels of expanded polystyrene (or styrofoam) that you fasten to the insides of windows at night. As drawing **79-A** shows, they are easy to make. Fasten them to windows with magnetic cabinet catches so they're easy to remove during daylight hours.

You may particularly want to install them in windows of rooms where you seldom need natural light (or in a few of the windows of a room).

79-A

magnetic base plates

7/8"

cabinet magnets

1"-thick polystyrene

Index

Photographers

Glenn M. Christiansen: 8, 16, 18
right, 50. **Sirlin Photographers:** 18
left. **Peter Whiteley:** 38.